A NON-MATHEMATICIAN'S GUIDE TO STATISTICS

AF787923

SECOND EDITION

Richard Flint
College of DuPage

cognella® | ACADEMIC PUBLISHING

Bassim Hamadeh, CEO and Publisher
Carrie Montoya, Manager, Revisions and Author Care
Kaela Martin, Project Editor
Christian Berk, Associate Production Editor
Miguel Macias, Senior Graphic Designer
Alexa Lucido, Licensing Coordinator
Natalie Piccotti, Senior Marketing Manager
Kassie Graves, Vice President of Editorial
Jamie Giganti, Director of Academic Publishing

Copyright © 2019 by Cognella, Inc. All rights reserved. No part of this publication may be reprinted, reproduced, transmitted, or utilized in any form or by any electronic, mechanical, or other means, now known or hereafter invented, including photocopying, microfilming, and recording, or in any information retrieval system without the written permission of Cognella, Inc. For inquiries regarding permissions, translations, foreign rights, audio rights, and any other forms of reproduction, please contact the Cognella Licensing Department at rights@cognella.com.

Trademark Notice: Product or corporate names may be trademarks or registered trademarks, and are used only for identification and explanation without intent to infringe.

Cover image copyright © Depositphotos/theromb.

Printed in the United States of America.

ISBN: 978-1-5165-2911-7 (pbk) / 978-1-5165-2912-4 (br)

cognella® ACADEMIC PUBLISHING

Contents

Preface

This is a wonderful time to study statistics. Apps make it inexpensive and convenient to collect and analyze data; social media firms, pollsters and physical scientists increasingly employ analytic and data mining techniques to make sense of data patterns; and digitization makes possible the collection of Big Data, immense collections of records that were hard to imagine only a few years ago. It seems that any field that relies on quantitative methods is experiencing a renewed interest in statistical methods.

Gone are the days when students needed advanced courses in mathematics to make sense of statistical methods. Statistics is still closely allied with mathematics, and the more calculus, linear algebra, ordinary differential equations, and numerical analysis, among others, the better one will understand the basic principles and results; but these subjects are no longer barriers to entry. Anyone who needs a data set described or graphed, or needs to run a statistical test, can easily find the routine and instructions, upload the data, and create a suitable presentation.

This book is written for the person who needs to use statistics, or is curious about its basic results and techniques, but does not care to invest years in building the appropriate theoretical framework; the person who wants to collect and analyze bivariate data, say, and needs the tools and definitions spelled out clearly, with examples and explanations.

This book is also written for students who, enrolled in a college or graduate-level course, forgot their basic statistics or never quite mastered all of it, or for some other reason needs a refresher. And it will be valuable for those in the general public who have never taken statistics and do not plan to, but are merely curious about what statistics is.

Having taught college mathematics since 1987, including quite a lot of statistics as well as calculus, computer programming, computer applications, and many other topics, I recognize the need for a book such as this one: easy to read but still covering the important topics while motivating them and explaining their important properties. Increasing numbers of students who have not taken the traditional route—four years of high school math, culminating eventually in Precalculus, or Calculus—now also need to take Beginning Statistics, or at least learn its main elements; but they don't always have the intuition that develops from years of doing mathematics, and anyway, many statistics textbooks are still written for people who have done math for years. Such books are fine, but this one is intended to fill the gap.

Although many colleges require completion of Precalculus or equivalent as a precondition, statistics actually uses very little algebra. If you forgot it, or lost your old algebra book, I provide a brief review of the algebra we will need in Appendix 1.

I've put the emphasis on discussion, demonstration, and illustration. The reader is encouraged to approach the text as a conversation, just as if (s)he were to chat with the instructor after class. If someone were to ask me "Why would someone prefer regression to ANOVA?" or "How do you actually use Bayes' Theorem?" or—as actually happened once with a colleague in the English department—"What is a standard deviation?" I would try to give a clear answer, without too much theory, using terms I hope (s)he would understand. I have also written this book in this manner.

One difficulty with math books—and, for that matter, courses—is that the terminology is introduced in progress, making it practically impossible to explain at the beginning what is to be covered. Maybe the following list will make more sense if you read the chapters first and then return to this preface, but here are some important features that make this book different from most other beginning statistics books:

- The concepts of normality, tests for normality, and normal distributions are introduced very early—in Chapter 1, in fact. Most users of statistics probably don't need to know that the normal density is $f(x) = \frac{1}{\sigma\sqrt{2\pi}}e^{-\frac{1}{2}\left(\frac{x-\mu}{\sigma}\right)^2}$. They probably want to know if the data are normal so they can justify using the Empirical Rule or, if the data fail normality, Chebyshev's Rule. I included a discussion of normality in the opening chapter on descriptive statistics because normality is an important enough concept to introduce almost immediately so it can be referenced throughout the book.

- Skewness and kurtosis work against normality. I included them here as part of the normality discussion.

- Also introduced early is the important procedure Single-Factor ANOVA (Analysis of Variance). Despite the complicated experimental design and calculations, ANOVA is fundamentally a hypothesis test, so I included it in that chapter immediately following the t distribution. Most books save ANOVA for the concluding chapters, which sometimes means that time runs out before it can be discussed carefully. In my courses, I devote about six to eight hours to it, finishing the discussion just before midterm. In my courses, and in this book, we cover Single-Factor ANOVA and Two-Factor ANOVA with replication and, in the regression chapter, Regression ANOVA.

- Bayes' Rule is discussed early in the chapter on Discrete Probability, immediately after defining probability. Many books do not discuss Bayes, which I think is a missed opportunity because Bayes is so important and so easy. My students never find it difficult; it takes about half an hour to illustrate two or three examples after establishing the concepts, definitions and notation of probability and conditional probability.

- Throughout, we rely heavily on Minitab and Excel. As I tell my students and have said once or twice in the book, a lot of beginning statistics is like changing the oil in our car: we should know how to do it, even if we usually pay others to do it. There's no getting around the fact that statistics has dozens of formulas, some of them very difficult for those without a mathematics background, and the only way to become familiar with the ideas is to work through the formulas. It remains true that most stats users click buttons to compute regression coefficients, test statistics, correlation coefficients, and the like, but if we're going to understand this subject, we need to have some understanding of what the software is doing. Nonetheless, the software is there as a useful check on our pencil-and-paper calculations. Furthermore, any aspiring user of statistics needs to practice using software since (s)he will spend so much time in front of a computer screen. Excel and Minitab are excellent for starters. Appendix 2 includes a list of the more useful Excel functions we will use in this book. More advanced users will probably use SPSS, SAS, Stata, or some other package.

- Concluding each chapter are supplementary materials that I wrote and actually administered to my students. I've found that those who come to my stats classes with recent algebra backgrounds expect the same kinds of homework exercises and style of tests, but I don't think the algebra paradigm translates well. In algebra, it's fine to introduce a topic like the Quadratic Formula, then assign a handful of drill problems and maybe choose one or two for quiz or test questions; but statistics is as much about the properties and limitations of the estimators as about the computations, and I believe the style of question should be similarly varied. It can take some getting used to, but is well worth it, and by the end of the course, most students appreciate the broadened perspective.

Chapter 1: Descriptive Statistics

Introduction

Our focus in this chapter will be, literally, how to describe a data set: what quantities will be of primary interest and how to compute or graph them. Minitab and Microsoft Excel will be used for many of our computations and almost all of our graphing.

First, we define and describe quantities that identify the center of a data set (mean, median, mode) and relative positions of the observations (percentile, quartile). Quartiles are particularly useful in the definition of outliers and in the construction of boxplots. Returning to measures of central position, we extend the discussion of the mean to account for observations with weights, such as in the computation of grade point averages.

Our analysis will be considerably extended by defining measures of dispersion, the most important of which is the standard deviation. Combined with the mean, the standard deviation describes the proportion of data in a specified interval in the distribution of interest. The most important such distribution is the normal, which we will discuss in some detail and with several examples. We will see that the area beneath the normal curve, over an interval on the x axis, can be represented as the probability of an event, provided we have the distribution's mean and standard deviation.

Measures of Centrality

We can divide statistics into two parts: descriptive and inferential. Descriptive statistics is concerned with the ways a dataset can be described, paying chief attention to measures of center, position, and shape. Inferential statistics deals with the way properties of a sample can be used to infer about the characteristics of an unknown population from which the sample is drawn, chiefly the population mean. Inferential statistics is far larger than descriptive, though it's useful to start with descriptive because it contains many terms, concepts, and techniques that we will use throughout. We will devote Chapter 2 to a deep discussion of sampling and inference procedures, but for the present, let us commence our study with descriptive statistics.

Suppose our dataset is a collection of reaction times to a certain drug. We will pretend that the following distribution is a population, not a sample; an unrealistic assumption, I suppose, but it relieves us of the burden of introducing issues like sampling and bias before we are ready to deal with them.

Drug reaction times in hours:

5	4	6	3	6.5	9	3	2	6	6	5	6.5	6

By far the most common question is: What is the center of the dataset? Surprisingly, "center" can have more than one meaning. We will begin with three interpretations: mean, median, and mode.

Mean

The mean of a dataset is obtained by adding the observations and dividing by the size of the dataset. In symbols, mean = $\frac{\sum x}{n}$, where Σ, throughout elementary statistics, is the summation operator. (Σ is the Greek letter capital sigma, σ is lowercase sigma.) The symbol X stands for our various observations. To be precise, we should be using the symbol X_i where the subscript i runs from the first to the last observation, that is, $X_1, X_2, \ldots X_{13}$; but this makes the notation messy without clarifying the underlying ideas, so I will usually avoid subscripts in this book. The number of observations in the data set, of course, is n.

For our dataset, mean = ΣX/n = 68/13 = 5.230769. We will follow convention and round to one more digit than the data were measured in and report mean = 5.23.

Use of the mean, more properly called the arithmetic mean, is ubiquitous. Students begin to "divide the sum of the values by n" in middle school to check that their teachers have issued the correct letter grades on their grade sheets. (A teacher who reports a C for someone whose arithmetic mean is above 90% can expect quite a lot of attention from the student and his or her parents.)

Second, practitioners of statistics almost always use software, not pencil and paper, to obtain such calculations. Since Microsoft Excel is so easy to use and widely available, we will focus mainly on it in this book, although we will occasionally use the statistics package Minitab. We enter the data as a column and choose a cell into which to type the command, in this case, **=average(a1:a13)** (Figure 1.1).

	A	B
1		5
2		4
3		6
4		3
5		6.5
6		9
7		3
8		2
9		6
10		6
11		5
12		6.5
13		6
14	mean	5.230769
15		

FIG. 1.1

As a side note, the word "average" technically can be used for any of the measures of centrality, not just the arithmetic mean. An average is any measure of central position. The popular conception, however, is that the mean is the average, and it seems that Excel follows this imprecise convention. We won't make a fuss about it. Note that in the above screenshot, I typed the label mean and then applied the =average() function to obtain the numerical result 5.230769. It's a good idea to label all your formula results, especially if you have many calculations on a spreadsheet.

Back to properties of the arithmetic mean. Did you observe that the value of the mean is not among the observations in the dataset? None of the drug reaction times equals 5.23 hours. It's unfortunate that, by this definition of center, there is no observation at the center. The median will correct this flaw, as we shall see.

Also, the mean is unstable with respect to increases in the maximum or decreases in the minimum values. This is also unfortunate, since we would prefer the center to be stable. To see this, we'll ask Excel to recompute the mean if our largest value, 9, were increased to, say, 50 (Figure 1.2):

	5
	4
	6
	3
	6.5
	50
	3
	2
	6
	6
	5
	6.5
	6
mean	8.384615

FIG. 1.2

The value of the mean follows the direction of increase (or decrease) in the extremes—or, as this is commonly expressed, "the mean is sensitive to outliers." If the max increases, so will the arithmetic mean. This is not a good statistical property—we don't like our centers bouncing around—although most of us instinctively know how to use it to our advantage. For example, if our course average going into the final exam is a high B, we surely would want to score as high on the final as possible in order to drag the mean upward.

To summarize, the mean is very well-known, but it is often not among the observations in a data set and is sensitive to outliers. The median is far superior, as we will see next.

Median

The median of a data set is the value in the middle position, assuming the data have been sorted in increasing (or decreasing) order. If there is an even number of observations, the median is the arithmetic mean of the two middle ones. Thus, for data sets with odd numbers of observations, the median overcomes both flaws we saw with the mean: the median often is one of the observations, and it is not sensitive to outliers. Even with an even number of observations, the median overcomes the sensitivity to outliers. The sorting does require extra work that the mean does not, but since this is done automatically by the software, we wouldn't notice. In the screenshot below (Figure 1.3), the Excel command for median is **=median(a1:a13).**

	5
	4
	6
	3
	6.5
	9
	3
	2
	6
	6
	5
	6.5
	6
mean	5.230769
median	6

FIG. 1.3

The median is based on position, not value. There are 13 observations in the above example; thus, the median will be observation 7, assuming the list has been sorted. It doesn't matter what the values are at the max or min, since only relative positions matter. That is why the median is not sensitive to outliers. Note that the mean has changed, although the median has not, in Fig. 1.4, although the maximum value has increased from 9 to 50.

Since it involves sorting rather than computation, there is no formula based on elementary arithmetic formulas for the median, as there is for the mean. For large data sets, we would use software.

	5
	4
	6
	3
	6.5
	50
	3
	2
	6
	6
	5
	6.5
	6
mean	8.384615
median	6
mode	6

FIG. 1.4

You can think of the median as the observation with the property that half of the distribution lie above it and half lie below it.

Mode

The mode is the easiest to explain: it is the value with the greatest frequency, or that which occurs most often. Be aware, though, that the mode doesn't always exist: there is no mode, for example, in the dataset {1, 2, 3, 4, 5, 6} because no observation occurs more often than any other. But let us return to our drug reaction times, and note that the Excel command for mode is **=mode().** As with the median, there is no formula based on elementary arithmetic operations for the mode. For large data sets, we would use software to find the mode, as with the median.

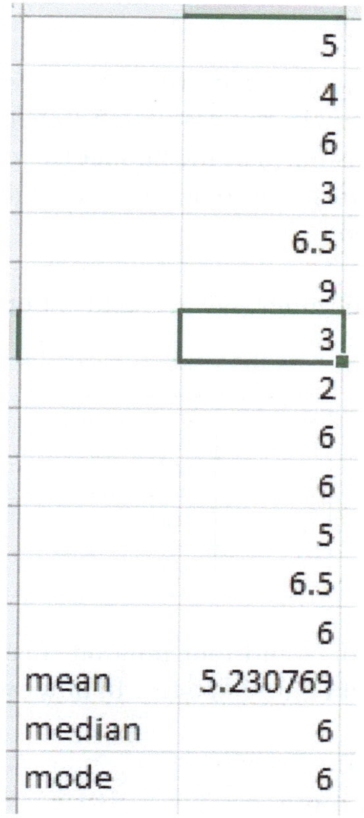

	5
	4
	6
	3
	6.5
	9
	3
	2
	6
	6
	5
	6.5
	6
mean	5.230769
median	6
mode	6

FIG. 1.5

If the mode doesn't exist, Excel reports the error code **#N/A,** "value not available."

Let us summarize the advantages and disadvantages of mean, median, and mode:

	Definition	Advantages	Drawbacks	Excel Command
Mean	sum of values divided by size of dataset	Taught and understood widely	1. Usually is not one of the observations 2. Sensitive to outliers	=average()
Median	value in middle position of sorted dataset	1. Usually is one of the observations 2. Not sensitive to outliers	Requires data to be sorted	=median()
Mode	value that occurs most often	Easiest to use	Does not always exist	=mode()

Weighted Mean

Mean, median, and mode are defined differently and can have different values, but we should not overemphasize the differences. In the case of an important type of data—the so-called bell-shaped distributions—it turns out that mean, median, and mode are all equal. More on this soon. For the moment, let us consider some examples where the mean is desired, but with weights applied to the observations.

Imagine a dataset with only three values, say 5, 7, and 8, but in which each value is repeated many times:

Value	Frequency
5	300
7	1000
8	450

Three hundred 5s, one thousand 7s and four hundred fifty 8s will be added and the sum divided by 1750, the total number of observations. In other words,

$$\text{mean} = \frac{5\times 300 + 7\times 1000 + 8\times 450}{300 + 1000 + 450}$$

$$= 5\times\frac{300}{1750} + 7\times\frac{1000}{1750} + 8\times\frac{450}{1750}$$

$$= 5\times 0.17 + 7\times 0.57 + 8\times 0.26 = 6.9$$

The answer 6.9 is still a mean, not a median or a mode, so it's sensitive to outliers (how would the mean change if the 450 observations of 8 were changed to 1000 observations of 8?) and is not to be found among the observations (no observation equals 6.9).

But the form of the solution is suggestive: the sum of products, in which each product is an observation times a decimal that represents a weight, and such that the sum of the weights equals 1. This form is very useful for a wide variety of important problems in which some observations count more than, or have greater weight than, some others. Let us look at two examples of weighted means: grade point averages and probability tables.

Grade Point Average (GPA)

Close to the heart of every degree-seeker, grade point average is an important example of a weighted mean. A student's grade point average (or GPA) is interpreted as his or her weighted mean performance throughout the term. The observations are the grades, and the weights are the proportions of credits each class is worth. Many colleges offer electives worth one or two credits; courses in the major are usually worth three or four credits. Thus, courses with more credits have the bigger weights, hence a bigger impact on GPA.

Most colleges quantify grades this way: A=4, B=3, C=2, D=1, F=0. Schools that additionally use plus/minus grading usually define "plus" to be 0.33 above the base grade and "minus" as 0.33 below the base grade, hence B+ = 3.33 and A- = 3.67. There is usually no A+. I've never seen or assigned grades of D- or F+, but I can't say they've never been used.

Course	Credits	Grade
Statistics	4	A
World War I	3	C
Epistemology	3	B
Renaissance Folk Music	3	C
Chemistry	4	B
Linux	1	C

The above student is taking 18 credits, so

$$GPA = 4/18(4) + 3/18(2) + 3/18(3) + 3/18(2) + 4/18(3) + 1/18(2) =$$

$$0.22(4) + 0.17(3) + 0.17(3) + 3/18(2) + 0.22(3) + 0.06(2) = 2.83.$$

The weights have been rounded in this text, of course, but not in the software. The unrounded weights do sum to 1, as they must. Thus, this student's mean performance is slightly weaker than a B, though not as low as B-.

Notice that a one-grade increase in the one-credit course Linux, from C to B, would increase the GPA to 2.89, but a one-grade increase in the four-credit course Chemistry increases the GPA to 3.06. This confirms our intuition that courses with more weight (or credits) have a greater impact on overall GPA.

Probability Tables

A closely related example is a probability table, in which the number of occurrences of an item of interest is matched with its probability, or proportion of outcomes, in the population. (For much of this book, the terms "probability" and "proportion" will be synonymous.) In the following probability table, let X represent the number of interviews an applicant goes through before being offered employment and p(X) represent the proportion of applicants who go through X interviews.

For example, 35% of applicants have one interview before being offered a position.

X	p(X)
0	0.07
1	0.35
2	0.54
3	0.04

We assume that the probability or proportion of job-seekers who interview four or more times is negligible. Also, note that the weights add to one, as they must in any distribution.

For probability tables, the weighted mean is given a special name: **expected value,** symbolized **EX.** Since X is the number of interviews, EX is the expected or "weighted average" number of interviews. To compute EX, multiply each value of X by its corresponding probability and then sum these products, exactly as with grade point averages:

$$EX = 0(0.07) + 1(0.35) + 2(0.54) + 3(0.04) = 1.55$$

Thus, the typical job-seeker has 1.55 interviews, or there are 155 interviews per 100 applicants, on average.

Measures of Relative Position: Max, Min, Percentiles, and Quartiles

Mean, median, and mode are the most common ways to describe the middle of a dataset, but it can be useful to go beyond the concept of centrality to include other measures of relative position.

The maximum is the biggest observation in a dataset, the minimum is the smallest. The range is the difference between them: range =max - min. Note that the range is a number, not a label. Especially for large datasets, it's a good idea to let the software pick out the largest and smallest values rather than trying to spot them ourselves. The corresponding Excel commands are =max() and =min().

The percentiles are 99 numbers that divide the data range into 100 intervals. A score in the kth percentile means that k% of the observations are smaller than the score in question. Results from standardized tests such as the SAT and GRE are often reported in percentiles for the following reason: Hundreds of thousands of students take such tests every year, and admission committees usually seek the top achievers. Transcript grades and recommendation letters are insufficient for the separation most admission committees desire, nor is it obvious how to compare grades across diverse institutions.

The percentile score can be high even if the raw score—the percentage of questions answered correctly—is small, and vice versa. Suppose your friend takes that GRE, and she returns that evening feeling proud that she answered at least 90% of the questions correctly in each section. But if all the other test-takers scored at least as well as your friend, her percentile score—the percentage of other test-takers whom she outscored—might be quite low. This measure of comparative performance is very important to admission committees. Conversely, suppose you have another friend who returns from the GRE stressed because he doubts that he answered half of the questions correctly. If he's lucky, the other test-takers will have scored even worse, and his percentile may be quite high. The percentile is a measure of relative position, not the percentage of questions answered correctly.

In Excel, the command for the kth percentile is **=percentile(range, 0.k).** If your data are in A1 through A500 and you desire the 80th percentile, you would enter **=percentile(a1:a500, 0.8).**

The opposite problem would be to input a value and call for its percentile rank. Excel provides the function**=percentrank(range, value)** where the list of data are in the column indicated by range, and the percentile rank of the value is required. If the data are in cells A1 through A500 and it is desired to know the percentile rank of, say, the value 71, you would enter **=percentrank(a1:a500, 71).**

Quartiles, Outliers, and Boxplots

Quartiles are conceptually similar to percentiles, except there are only three quartiles that divide the data range into four intervals. In increasing order, these are Q_1, Q_2, and Q_3. You can think of Q_1 as the median of the bottom half and Q_3 as the median of the upper half, while the second quartile, Q_2, equals the overall median, which also equals the 50th percentile: $Q_2 = P_{50}$ = median. The interquartile range, IQR, is the difference between Q_3 and Q_1, or $IQR = Q_3 - Q_1$.

Quartiles are the basis for the statistical definition of outliers. In ordinary language, we think of an outlier as an unusually large or unusually small observation. If a Fortune 500 CEO were to move into a college dorm, say for some marketing or publicity purpose, she would probably be unusually wealthy relative to her neighbors, hence an upper outlier relative to her personal wealth; if an NBA player were to do the same, he would probably be unusually tall relative to his neighbors, hence an upper outlier relative to height.

Statistically, we make this precise with the following definition:

An **upper outlier** is an observation that exceeds Q_3 by more than 1.5 times the IQR.
A **lower outlier** is an observation that is less than Q_1 by more than1.5 times the IQR.

In searching for upper outliers, we first establish the upper outlier boundary $Q_3 + 1.5 \times IQR$, then examine whether there are observations above this boundary. If there are, they are outliers. Keep in mind that this threshold does not necessarily correspond to an observation; in fact, the threshold may be bigger than the maximum or lower than the minimum, which of course is impossible for any actual observation.

The definition is similar for the lower half of the dataset: the threshold for lower outliers is $Q_1 - 1.5 \times IQR$. Any observation below this threshold is a lower outlier, although of course this threshold may not itself be an observation.

The presence of outliers is important in the discussion of normal distributions: one of the chief reasons that distributions fail normality is the presence of outliers in the distribution. We will have quite a lot to say about normality later in this chapter.

Suppose you have the following summary of a dataset. Are there any upper outliers?

Max	110
P_{90}	90.5
Q_3	80
Q_1	25.5
Min	23.0

First, calculate the boundaries. The upper outlier threshold is

$$Q_3 + 1.5 \times IQR = 80 + 1.5 \times (80 - 25.5) = 161.75.$$

Since this threshold exceeds the max, we know that there are no upper outliers. As an exercise, convince yourself that, since the lower outlier boundary is lower than the minimum, there can be no lower outliers.

Quartiles are also essential to the graphical technique called the boxplot. The three quartiles, in combination with the max and the min—the so-called 5-number summary—are arranged in order, with a box drawn around the first and third quartiles. Minitab has a very easy-to-use menu option for boxplots, as shown in Fig. 1.6.

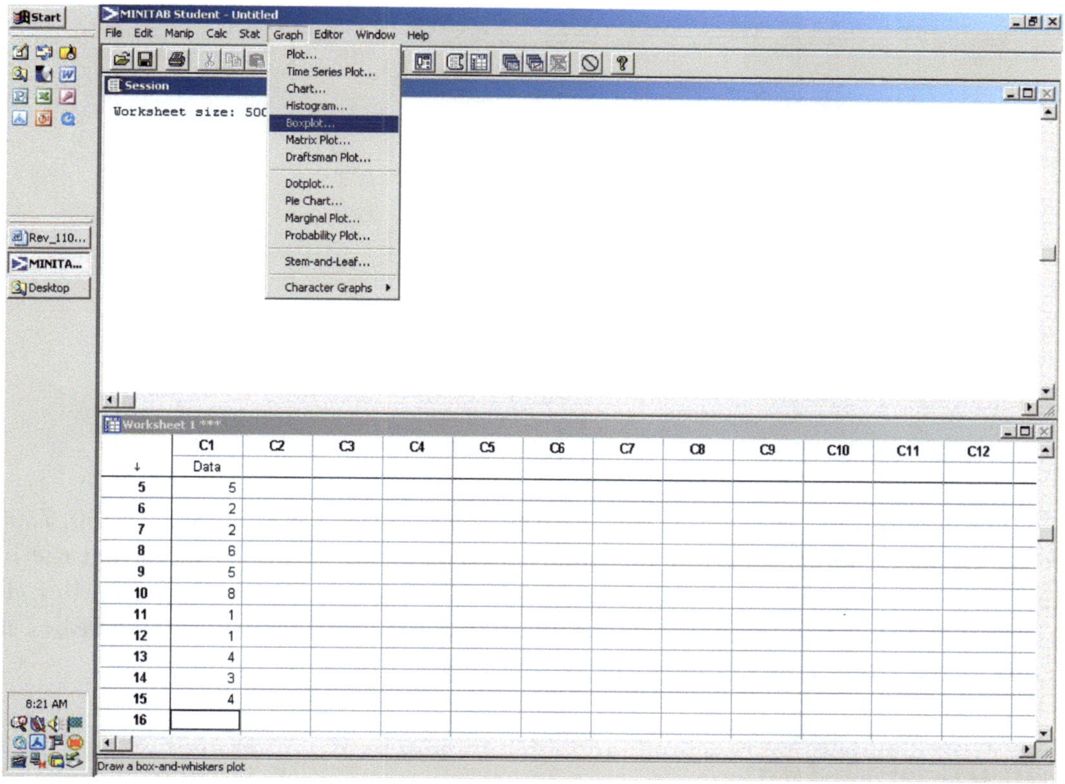

FIG.1.6

The following boxplot is based on a dataset of 123 observations whose 5-number summary is {min, Q1, Q2, Q3, max} = {1,57,73,87,113}. The asterisk at the bottom indicates a lower outlier, apparently the minimum observation 1.

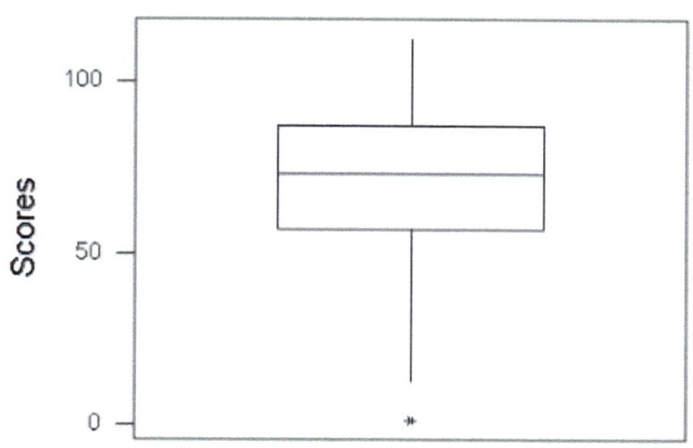

FIG. 1.7

Boxplots are useful because the box surrounding Q_3 and Q_1 represents "the bulk of the data." (The height of the box equals the IQR; the width is unimportant.) This can generate interesting research questions. For example, suppose the following boxplot represents income data for a certain Chicago neighborhood:

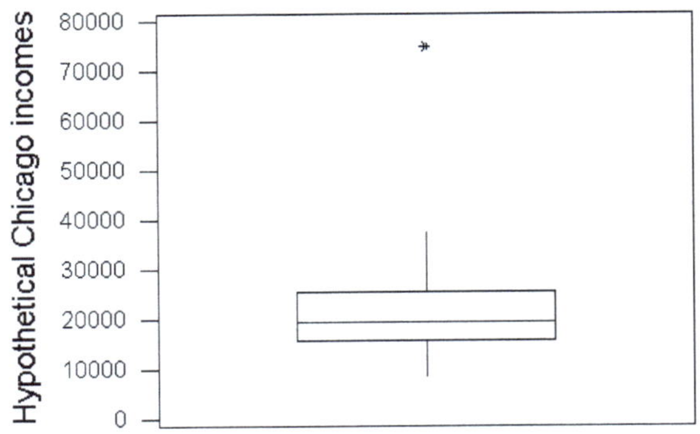

FIG. 1.8

The 5-number summary for this distribution is {8215, 15681, 19518, 25330, 75000}. While we wouldn't want to hazard any premature conclusions, it does seem that the bulk of the data are clustered near the minimum, with the maximum as an upper outlier. To a social scientist, this may suggest research hypotheses involving poverty-related issues such as quality of schools, scarcity of jobs, and desirability of welfare programs.

Standard Deviation

Measurement of relative position is a useful way to begin analysis of a dataset, but this can be made immensely more productive if we include a measure of spread of the data around its mean.

Imagine the average price of gas in your area is \$2.50/gallon. This could mean that all stations are charging \$2.50. If this were the case, you may as well fill up at the next station on your route—there would be no advantage to driving further in search of a lower price. On the other hand, it could be the case that half the stations are asking \$3.50 and half \$1.50. Or maybe half the stations are offering \$4.50 and the other half \$0.50. In all cases, the mean price would still be \$2.50, but there would be a great advantage in driving in search of the lower price.

The point is that the mean is unable to distinguish the spread, or dispersion, of the observations around their center. There is more than one way to quantify spread, but by far the most useful is the standard deviation.

Formulas for the Standard Deviation

The Definitional Formula

The standard deviation of a data set is the square root of the average of the squared deviations of each observation from the mean of the data set.

There are a lot of ideas in this definition; we will start the idea of "squared deviations from the mean." Compute the mean of the data set, then square the difference between each observation and the mean. Average these squared differences, and the square root of this average is the standard deviation.

$$\text{population standard deviation} = \sqrt{\frac{\sum (x - mean)^2}{n}}$$

In the above formula, the symbol x ranges over the observations in the data set, n is the number of observations, and Σ is the summation symbol.

Some readers may recognize this formula with the Greek letters μ for population mean and σ for population standard deviation. We will motivate and discuss this notation in the next chapter, when we distinguish between population parameters and sample statistics, but there's no harm in peeking ahead:

$$\sigma = \sqrt{\frac{\sum (x - \mu)^2}{n}}$$

In practice, standard deviations are computed far more often for samples than for populations. In such cases, we would replace σ with s, the sample standard deviation, and the n in the denominator is replaced by n-1:

$$s = \sqrt{\frac{\sum (x - \bar{x})^2}{n - 1}}$$

where the symbol $\bar{x}$ refers to the mean of a sample. For the remainder of this chapter, though, let us analyze the formula for the population standard deviation.

Example

The six observations in your sample are 1, 2, 4, 4, 8, 11. The mean is 5, so the squared deviations from the mean are

$(1-5)^2 = 16$, $(2-5)^2 = 9$, $(4-5)^2 = 1$, $(4-5)^2 = 1$, $(8-5)^2 = 9$, $(11-5)^2 = 36$.

The average of these squared deviations is (16+9+1+1+9+36)/6 = 12, so the population standard deviation σ is $\sqrt{12} = 3.5$, rounded.

A few interpretive comments are in order. First, remember that we have computed the standard deviation of a population, not a sample. This is rarely done, actually; sample standard deviations are far more common since samples are far more easily available than populations. But sampling opens up a considerable number of other issues, and we can't discuss everything at once, so we will table the discussion of sample standard deviations for the time being.

Second, we might ask what, exactly, is the quantity that corresponds to our computation? In the case of measures of centrality, the answer is obvious: $\Sigma X/n$ is the center of the dataset. But what is "square root of the average of the squared deviations from the center"? The rather unsatisfying answer is, the spread. The concept of spread is, shall we say, diffuse; it doesn't have a more concrete interpretation. All we can say is first, the bigger the standard deviation, the greater the spread around the mean; and second, often in the field of statistics, smaller standard deviations are preferable. We will discuss this further when we come to a discussion of confidence intervals for population means in the next chapter.

Third, in the computation of the standard deviation, why do we average the squared deviations from the mean instead of, say, taking the absolute value of the deviations from the mean? Let's answer this in two parts. First, if we considered only the deviations from the mean with no further operations, our standard deviations would always equal zero because the mean is the center, hence the half of the observations below would balance the half above. In the previous example, the deviations are -4, -3, -1, -1, 3, and 6, whose sum is exactly 0.

Why, then, do we choose to square the differences rather than some other operation whose results are always positive? The answer to this concern is a little complicated and beyond the scope of this text, but it is based on the fact that almost all the formulas throughout basic statistics result from certain operations from differential and integral calculus. That is, it results from certain deeper mathematical assumptions and operations. The complete justification for the formula can be found in books on mathematical statistics.

Next, we investigate the purpose of the square root—the final step in the computation. The answer involves the units in which the data were measured. Suppose, in the above example, the original observations were measured in feet. Then the mean would also be measured in feet. But each squared deviation would be in square feet, hence so would be the mean of the squared deviations. The purpose of the square root is to return the standard deviation to the units in which the data were actually measured, that is, to cancel the square.

A closely related quantity that also measures dispersion around the mean is called the variance, which is the square of the standard deviation. Equivalently, the variance is the quantity beneath the square root sign. In the example, the variance is 12 and the standard deviation is $\sqrt{12}$ or 3.5, rounded. Many results in statistics are stated in terms of variances instead of standard deviations, but these are not radically different concepts. A synonym for variance is mean square, literally the mean of a set of squared differences; we will have quite a lot to say about mean squares in the ANOVA section in the next chapter.

As you might expect, in practice people rarely compute standard deviations by paper and pencil; instead, they use software. I always ask my students to practice the definitional formula because, like changing our oil, it's the sort of thing we should know how to do even if we don't always do it ourselves. In Excel, if the data are in cells A1 to A100, the formula is **=stdevp(A1:A100)**, the trailing **p** reminding us that we have computed the population parameter. If the more common sample standard deviation is required—the formula with n-1 replacing n—the Excel formula is **=stdev(A1:A100).**

The definitional formula for the standard deviation has the advantage of clarity, which is why I ask my students to become familiar with it. The ability to use the definitional formula for computing standard deviations contributes to an understanding of what a standard deviation actually is. The disadvantage is that it takes so many steps. Historically, textbooks also featured this much more convenient shortcut formula:

$$\sigma = \sqrt{\frac{n\sum x^2 - (\sum x)^2}{n^2}}$$

It's a shortcut because it does not involve the population mean μ, hence there is no need to compute squared mean differences $(x-\mu)^2$. We need only square the observations and sum them; then sum the observations and square this quantity; and multiply, subtract, and square root as indicated. Using the quantities from the above example,

$$\sum x^2 = 1+4+16+16+64+121 = 222$$

$$(\sum x)^2 = (1+2+4+4+8+11)^2 = 900$$

$\sigma = \sqrt{\frac{6\times 222 - 900}{36}}$ = 3.5, rounded, as we expected.

The corresponding shortcut formula for s is $s = \sqrt{\frac{n\sum x^2 - (\sum x)^2}{n(n-1)}}$

As a side note, remember the difference between Σx^2 and $(\Sigma x)^2$. The first means "square each term, then sum the squares," while the second means "sum the terms, then square that total." See Appendix 1 if you want further discussion and practice with order of operations.

Probability tables use formulas for the standard deviation that look somewhat different. Recall that the symbol EX is the weighted mean of the random variable X, where the weights are the probabilities or proportions of X. EX is the same quantity as μ. Similarly, EX^2 is the weighted mean of X^2: square each value X, then multiply that square by the weight. Then the variance of X is

$\text{Var } X = EX^2 - (EX)^2$, or $EX^2 - \mu^2$

and the standard deviation of is stdev(X) = ...

For the above probability table

X	p(X)
0	0.07
1	0.35
2	0.54
3	0.04

we already computed $\mu = EX = 1.55$. For the standard deviation, we first need EX^2, which we compute by squaring each value of X, multiplying each square by the corresponding probability of X, and summing these products:

X	0	1	2	3
X^2	0	1	4	9
p(X)	0.07	0.35	0.54	0.04

$$EX^2 = 0\times0.17 + 1\times0.35 + 4\times0.54 + 9\times0.04 = 2.87$$

Thus, $\text{Var } X = EX^2 - \mu^2 = 2.87 - 1.55^2 = 0.4675$, and $\sigma = \sqrt{0.4675} = 0.68$, rounded.

Descriptive Statistics Using the TI-83

Your statistics class might call for a graphing calculator such as the TI-83. This will enable you to report univariate statistics, bivariate statistics, simple hypothesis tests and regression information. More about that later, but for now, let's use it to report univariate data:

Turn the calculator on, then create your data list by choosing Stat – Edit, then option 1 (Edit). Enter your data in the column labeled L1. Then press Stat, choose Calc and option 1 (1-variable statistics). You will be rewarded with the sample mean, the sum of the values, the sum of squares, the sample and population standard deviations, the number of observations, and the five-number summary.

Let us conclude this section with two final observations: First, remember that the only people who compute standard deviation by hand are statistics students, and this is to enable them to become familiar with its deeper meaning. Everyone else uses software. Second, remember that the standard deviation, by itself, does not tell us much that is useful about the data. What is extremely useful is the combination of σ with the population mean μ, and it is to this discussion that we now turn.

Empirical Rule

The standard deviation of a data set is essential because, in combination with the mean, it quantifies the proportion of data within a specified range around the mean. Our first result, the Empirical Rule, describes how this works for bell-shaped data:

For bell-shaped data,

- about 68% of the observations are within one standard deviation of the mean;
- about 95% of the observations are within two standard deviations of the mean;
- about 99% of the observations are within three standard deviations of the mean.

Before we turn to examples, let's say a word about bell-shaped data. Perhaps the easiest way to understand it is as the limiting case of a histogram:

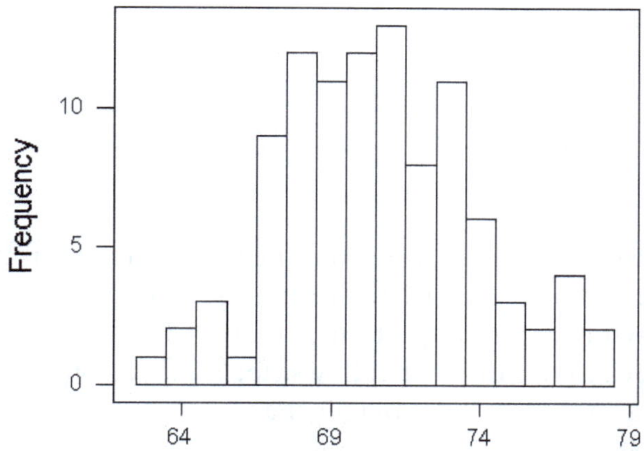

FIG. 1.9

Figure 1.9 above, which was created with Minitab, is the graph of data that are nearly, but not perfectly, bell-shaped. This is the sort of data to which the Empirical Rule applies. Basically, data are bell-shaped if the histogram represents a bell: low frequency at the far left and far right, increasing toward the mode, and symmetric. Normality is one very important type of bell-shaped data, about which we shall have much more to say shortly. For the moment, let us record this important fact: ***for bell-shaped data, the mean, median, and mode are all equal.***

Suppose we have bell-shaped data, with $\mu = 80$, $\sigma = 15$. By the Empirical Rule,

68% of the observations are in the interval (80 – 15, 80 + 15) or (65, 95);
95% of the observations are in the interval (80 – 2(15), 80 + 2(15)) or (50, 110);
99% of the observations are in the interval (80 – 3(15), 80 + 3(15)) or (35, 125).

According to the Empirical Rule, the combination of mean and standard deviation can be used to describe the proportion of data in intervals centered at the mean, where the proportion is a function of the number of standard deviations. An even more specific result is available, which we will study in the next chapter, but it requires a little more knowledge about the distribution than its bell shape. Still, as a back-of-the-envelope type of calculation, the Empirical Rule can be very useful.

Chebyshev's Rule

Suppose we desire some estimate on the proportion of observations around the mean, but the distribution is not bell-shaped:

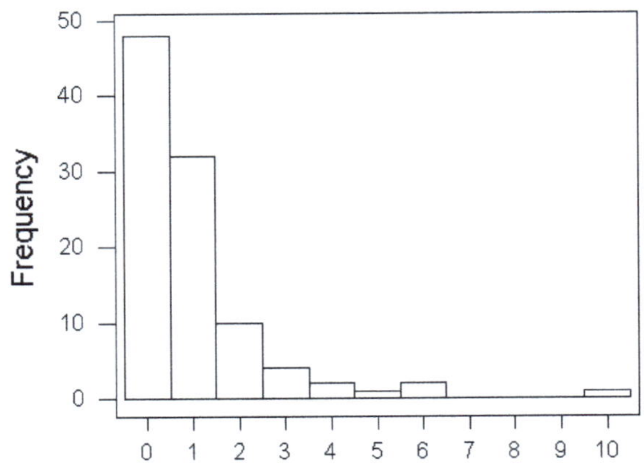

FIG. 1.10

Unfortunately, there is no exact result corresponding to the Empirical Rule for skewed data such as in Fig 1.9. There is, however, a partial result due to Pafnuty Chebyshev; it is partial because, rather than giving exact proportions, it gives lower bounds corresponding to numbers of standard deviations. In what follows, let k stand for the number of standard deviations around the mean, letting k be bigger than 1:

> **Chebyshev's Rule** The proportion of data within k standard deviations of the mean μ is never less than $1-1/k^2$

In symbols, $p(|X-\mu| \leq k\sigma) \geq 1-1/k^2$.

Suppose our non-bell-shaped distribution has mean 49, standard deviation 11. The Empirical Rule doesn't apply, but by Chebyshev, the proportion of observations within, say, two standard deviations of 49 is at least 75%, and the proportion within three standard deviations of 49 is at least 89%.

Normal Distributions

An extremely important type of distribution is the normal. Technically, there is a different normal distribution for each μ and σ although such distributions can be reduced to a common form, or standardized. We won't say much about standardizing normal curves here; what is important is that curves of all normal distributions have a characteristic bell shape:

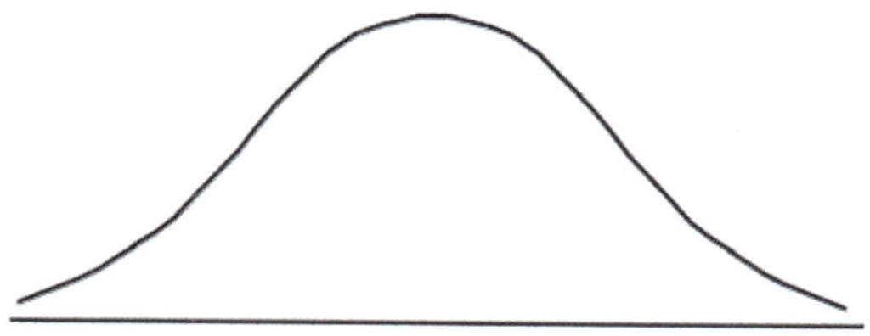

FIG. 1.11

It's not that normal distributions occur very often in the real world; they do not, although we will see in the next chapter a type of data that is always normal, at least asymptotically: means of samples. What makes the normal distribution so important is that it can approximate other extremely complicated or otherwise unpleasant distributions, particularly the t distribution, the binomial when the number of trials, the number of successes, and the number of failures are all big. For this reason, it's important to have a firm understanding of the properties of the curve of an idealized normal distribution.

Properties of the Normal Curve

1. The total area beneath the curve equals 1.
2. As for all bell-shaped curves, mean=median=mode. The mode is the value on the x axis that corresponds to the curve's peak.
3. The curve is symmetric about its mode.
4. The x axis is an asymptote, hence there are no x intercepts.
5. Areas beneath the curve are interpreted as areas as long as the values on the x axis have been transformed into their z scores.

The "1" that represents the area beneath the normal curve has no units since it is a probability. Point 2 can be a stumbling block for students, who sometimes confuse points on the x axis with the areas above these points. Point 3 implies that each half of the curve has area 0.50; it also implies that the area to the right of a positive x value, say x_0, equals the area to the left of $-x_0$. This sometimes considerably simplifies our work.

Regarding point 4, recall from College Algebra that an asymptote is a line that a curve approaches increasingly closely but does not touch, in the limit as one of the variables approaches a particular value (or grows without bound). What point 4 expresses is that the normal curve does not touch the x axis. Statistically, this means that there is no observation such that the probability of getting a more extreme outcome is 0; there is always the probability, however small, of getting an outcome that is bigger in absolute value than the outcome you actually got.

Point 5 introduces the concept of the z score. It is very important to transform our observations into their z scores, or else the probability interpretation of area won't work. Therefore, let us take a short detour to explain what a z score is and how it is computed.

z Score

The z score of an observation measures its distance from the mean, where the unit of measurement is the standard deviation.

Suppose we have a distribution with mean 70, standard deviation 3. Then the observation 73 is three units bigger than the mean but only one standard deviation bigger, hence its z score equals 1:

$$z_{73} = \frac{(73-70)}{3} = 1$$

For this same distribution, the mean 70 has a z score equal to 0 because it is 0 standard deviations away from itself:

$$z_{70} = \frac{(70-70)}{3} = 0$$

It is always the case that the mean of a distribution has a z score equal to 0.

If an observation is below the mean, it will have a negative z score. The observation 64, for example, is two standard deviations below the mean, so its z score is -2:

$$z_{64} = \frac{(64-70)}{3} = \frac{-6}{3} = -2$$

Z scores can be negative, positive, zero, or any decimal. In the context of normal distributions, z scores greater than about 3.5 in absolute value—greater than 3.5 or less than -3.5—are quite rare, and soon we will see why.

The z score formula we have been using can be written

$$z_x = \frac{x-\mu}{\sigma}$$

Note that in the numerator, the observation is written before the minus sign, the mean after it. Also keep in mind that, to this point, we are still dealing with populations, not samples.

Z scores are important in the analysis of normally-distributed data because there exist tables and functions that match the observation's z score with the area to its left; with Excel, use the norminv() function.

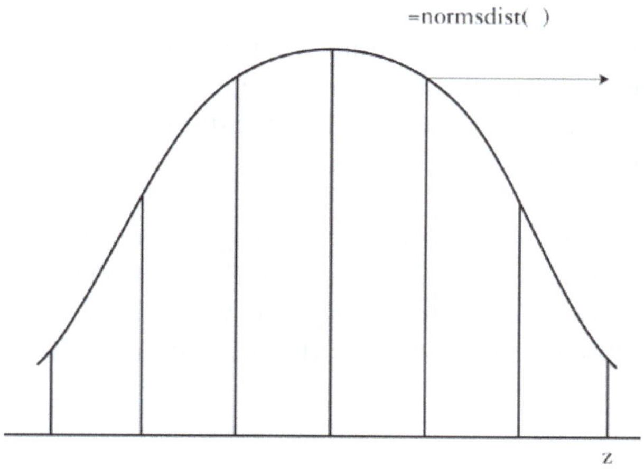

FIG. 1.12

To summarize: Normally-distributed data have the very nice property that probability questions can be interpreted as areas beneath the normal curve. To do this, we first transform the observation into its z score, then use software or a normal distribution table to match the z score to the appropriate area. If we desire a left-tail probability, our problem will be solved; otherwise, it's easy to make suitable modifications.

Example

Employee commute time at a certain large company is normally distributed with $\mu = 35$ min, $\sigma = 5$ min. X will represent the commute time of a randomly chosen employee. Let us compute $p(X < 30)$, the proportion of employees whose commute is less than 30 min—equivalently, the probability that $X < 30$. (Throughout this book, we will use the terms "probability" and "proportion" synonymously, unless otherwise indicated.) First, we draw a bell-shaped curve indicating the relevant observations and their relationship to the mean (or median, or mode):

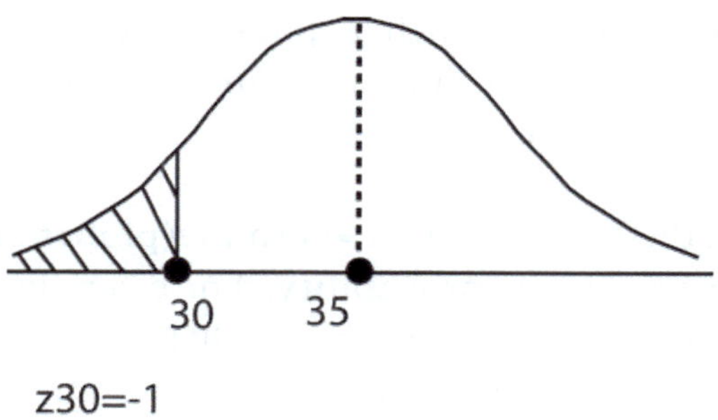

FIG. 1.13

The z score for 30 is $z_{30} = \dfrac{30-35}{5} = -1$. (Since the observation 30 is less than the mean, its z score is negative.) The Excel function that reports left-tail areas is **=normsdist(z)** where z is the s score of the observation in question. In our case, **=normsdist(-1)** = 0.1587, rounded. Thus, $p(X < 30) = 0.1587$, or about 15.9% of employees' commute time is less than 30 minutes.

Because the normal distribution is continuous rather than discrete, probabilities are not changed by the presence or absence of an equal sign: $p(X < 30) = p(X \leq 30)$ and similarly for problems involving > and ≥. However, when we discuss distributions in Chapter 4, we will need to start paying attention to equal signs.

Let us return to our example. How can we answer questions about right-tail probabilities, such as $p(X > 38)$, using the =normsdist() function? The answer is to be found in property 1: the total area beneath the normal curve equals 1. We will use Excel to compute the left-tail probability, then subtract it from 1 to get the desired probability:

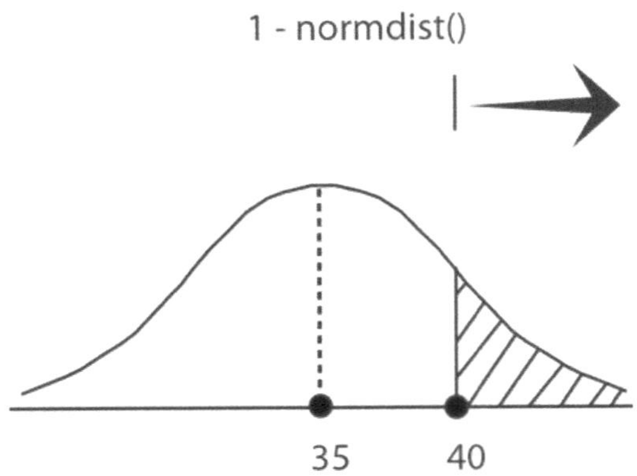

FIG. 1.14

$p(X > 40) = 1 - \text{normsdist}(-1) = 0.2742$. Approximately 27.4% of employees' work commute takes longer than 38 minutes.

Z scores can, in principle, have any value. But when used with normal populations, z scores whose absolute values are in excess of 3.5 are rare. This might be surprising at first, but it makes sense when you think about it: the left and right tails become indistinguishable from the x axis, so there must be some point beyond which we hardly expect to see any observations. The Empirical Rule, which applies to bell-shaped data, does say that approximately 99% of data are within three standard deviations of the mean; the normal distribution function makes this more precise.

Consider this experiment: the Excel function **=norminv(probability, μ, σ)** returns x values such that the area to the left of x is **probability**. For example, =norminv(0.0001, 35, 5) returns 16.40491757, because the tail to the left of 16.40491757 is 0.0001, assuming a normal distribution with mean 35 and standard deviation 5.

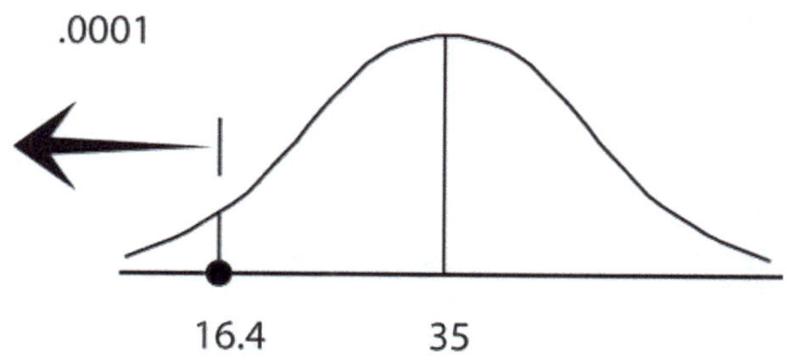

FIG. 1.15

Note that $z_{16.40491757}$ equals approximately -3.72, confirming that the tail areas defined by observations roughly ±3.5 standard deviations away from the mean (in this case, 3.72 standard deviations to the left of μ) are, practically speaking, zero. Try modifying this command by varying the probability, then computing the z scores of the resulting values. Convince yourself that z scores become smaller in absolute value as the probabilities increase, and vice versa. Sketch a curve for each case to help clarify the geometry.

To return to our employee data, let X represent a commute time of less than 10 minutes. Since z_{10} = -5, then p(X < 10) = normsdist(-5) = 2.87E-07, which equals 0.000(rounded). The proportion of employees who commute less than 10 minutes is vanishingly small. This is not surprising when we realize that the observation 10 is five standard deviations below the mean. Since approximately 99.9% of observations are within ±3.5 standard deviations of μ, we expected an observation five standard deviations away from the mean to have very low probability.

What proportion of employees commute between, say, 40 and 45 minutes? In this case, the required area, p(40 < X < 45), is neither a left nor right tail, but rather a chunk between two endpoints. The key is that the required area is the difference between two left-tail areas:

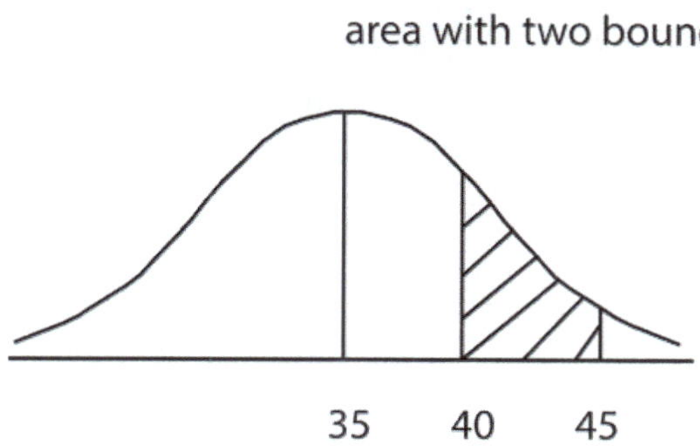

FIG. 1.16

The bigger area is =normsdist(2) = 0.9772; the smaller is =normsdist(1) = 0.8414. Hence, the required area is 0.9772 – 0.8414 = 0.1359, rounded of course.

Probabilities are always between 0 and 1, inclusive. But now that we have started doing arithmetic with probabilities, there arises the possibility that our calculations will contain an error, and we may mistakenly report probabilities that are less than 0 or greater than 1. Always be alert for such mistakes. Unfortunately, there is no way to check our probability answers by plugging answers into original equations, as in algebra. Nothing is more embarrassing than reporting something like "p(40 < X < 45) = -0.1359" because we subtracted terms in the wrong order.

Speaking of common errors, it is important to remember never to combine z scores: never add them, subtract them, or do anything except associate them with left-tail areas. Areas can be subtracted from each other, or subtracted from 1, as the problem requires, but never z scores. In the previous example, note that we found the area to the left of 45 to be 0.9772 ($z_{45} = \frac{45-35}{5} = 2$) and the area to the left of 40 to be 0.8413 ($z_{40} = \frac{40-35}{5} = 1$), then subtracted the smaller area from the bigger; but we emphatically did not start by subtracting the smaller observation from the bigger, or the smaller z score from the bigger. Some students think this would be an acceptable shortcut, but it isn't.

Minitab's Normality Tests

We've had so much to say about normality that you may have wondered how it could be determined, with an actual data set, how to assess it. Surely it would not be sufficient to base the determination on a histogram. Furthermore, the quality of normality is robust, meaning that small departures from perfect normality are acceptable in practice.

One answer is provided by the normality test option in Minitab. The data are hypothesized to be normal, a test statistic is computed—Minitab spares us the details—and the result is reported by way of a quantity called the p value: if p is greater than 0.05, the data are normal, but if p is less than 0.05, they are not. We will dedicate most of Chapter 2 to the subject of hypothesis testing and p values so we've gotten somewhat ahead of ourselves here, but occasionally we need to make use of a concept before we've completely explained it.

In the following screenshot, I've performed a normality test on 100 observations from a normal distribution with $\mu=0$, $\sigma=1$. Not surprisingly, the p value for this distribution is quite high: not merely greater than 0.05, but actually greater than 0.50 (being a probability, the p value is always between 0 and 1).

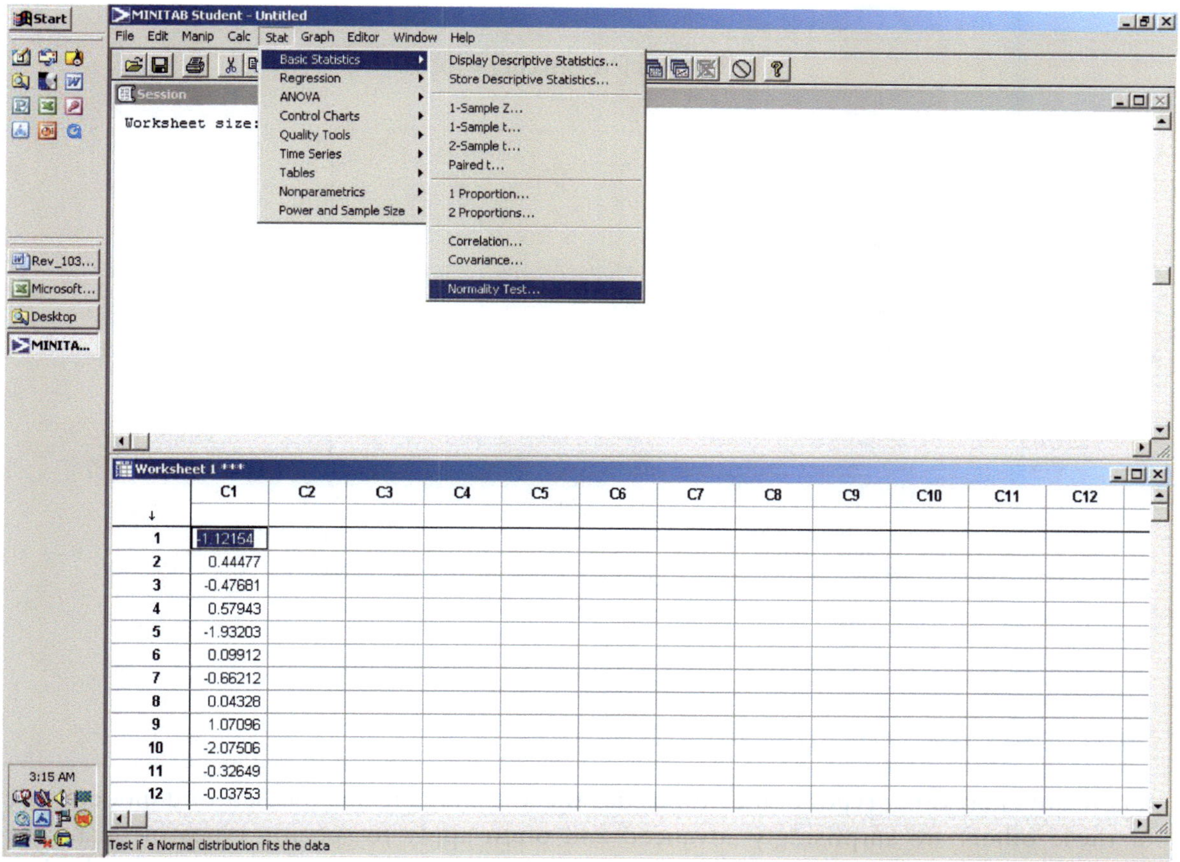

FIG. 1.17

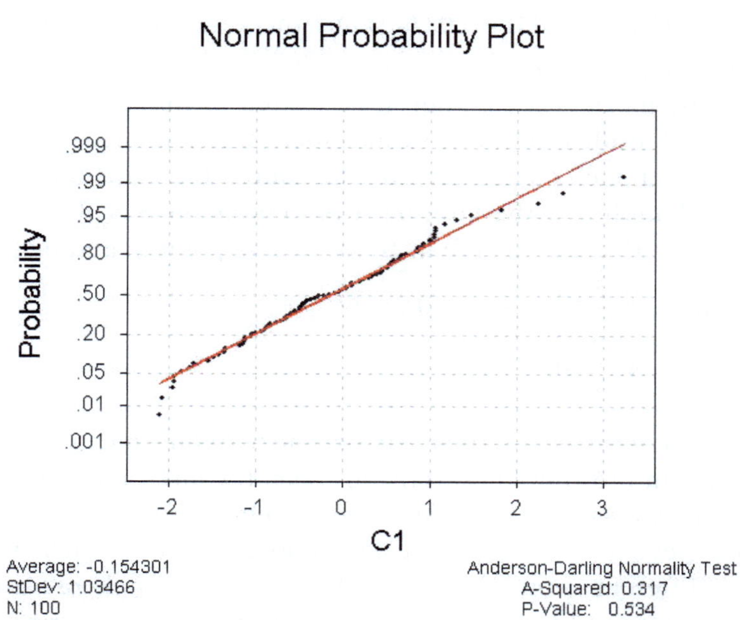

FIG. 1.18

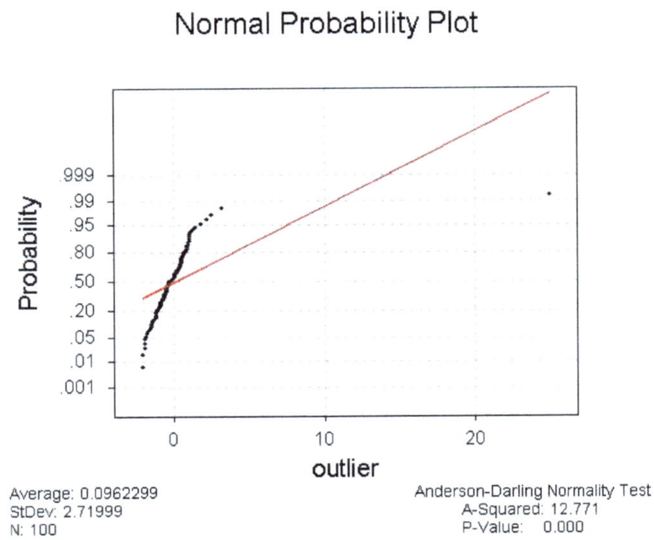

FIG. 1.19

Suggestions for Further Reading

There are many other types of means besides the arithmetic mean, median, and mode. Some of these others are surprisingly advanced and often apply to specific types of data. The geometric mean, defined as the nth root of the product of n percentages, is used when analyzing annual percent increases, say, in a firm's profits. If Year 2 profit over Year 1 is 2.5%, Year 3 profit over Year 2 is 1.5%, and Year 4 profit over Year 3 is 3.0%, then the geometric mean equals $\sqrt[3]{(1.025)(1.015)(1.03)} = 1.0233$. Thus, the single profit that could be substituted for the various actual ones and yield the same overall result would be 2.33% annual profit.

The harmonic mean is used when it is desired to find the one single speed that yields the same overall result as multiple various speeds, assuming roads of the same length. It is the number of data points, n, divided by the sum of their reciprocals. Say roads A, B, and C are the same length, but the limit on A is 45 mph, on B is 30 mph, and on C is 55 mph. The single speed that would take us from starting point to ending point in the same amount of time equals $\dfrac{3}{\dfrac{1}{45}+\dfrac{1}{30}+\dfrac{1}{55}} = 40.7$ mph.

Note that neither the geometric mean nor the harmonic mean equals the arithmetic mean. There are very interesting and surprising geometric interpretations and relationships among the three; the *Mathematical Gazette* is a good source for articles.

There are several variations on the normality test, each with a different methodology and different properties. Minitab allows you to choose among the Ryan-Joiner, Anderson-Darling and Kolmogorov-Smirnov tests; the second two are based on the empirical cumulative distribution function, while Ryan-Joiner is basically a regression. See

Minitab's support page, support.minitab.com, for a bit more detail. Thode's book is a good place to start.

Expected value is an extremely important statistical concept. One of its uses is in the definition of bias. We will discuss bias in sample surveys in the next chapter, transformations to normality, but at this early point we can say that the bias of an estimator is the difference between its expected (average) value and its true value. An estimator is said to be unbiased if the difference equals zero. There is a tradeoff between bias and variance; sometimes it's good to live with a little bias if the alternative is unacceptable variance. Kennedy has more details.

References

Balanda, Kevin P. and H. L. MacGillivray (1988). "Kurtosis: A Critical Review," *The American Statistician*, 42, 111-119.

Boslaugh, S. and Paul Andrew Watters (2008). *Statistics in a Nutshell*. Sebastopol, CA: O'Reilly.

Di Domenico, A. (2005). "The Right Triangle and the arithmetic, geometric and harmonic means." *Mathematical Gazette*, 89, 261.

Kabayashi, Yukio (2002). "Proof Without Words." *Mathematical Gazette*, 86:2.

Kennedy, P. (1994). *A Guide to Econometrics*. Cambridge, MA: MIT Press.

Macdonald, Matthew (2007). *Excel 2007: The Missing Manual*. Sebastopol, CA: O'Reilly.

McKenzie, J., Robert L. Schaeter, and Elizabeth Farber (1995). *The Student Edition of Minitab for Windows*. Reading: Addison-Wesley.

McClave, James T. and Terry Sincich (2009). *Statistics*, 11th ed. Upper Saddle River: Pearson.

Moore, David S. and George P. McCabe (1993). *Introduction to the Practice of Statistics*. New York: W. H. Freeman and Company.

Ruppert, D. (1987). "What Is Kurtosis?" *The American Statistician*, 41, 1-5.

Snedecor, George W. and William G. Cochran (1989). *Statistical Methods*, 8th ed. Ames: Iowa State University Press.

Thode, H. (2002). *Testing for Normality*. Marcel Dekker.

Tukey, John (1977). *Exploratory Data Analysis*. Reading: Addison-Wesley.

Urmanin, Z. (2002). "A Generalization of the Arithmetic Mean: Geometric Mean: Harmonic Mean Inequality." *Mathematical Gazette*, 86, 293-296.

Chapter 1 Glossary

5-number summary	From lowest to highest, these are Min, Q_1, Q_2, Q_3, Max. Used in the construction of boxplots.
boxplot	Graphical technique, based on the 5-number summary, for displaying a data set. It indicates outliers and a box that surrounds the interquartile range.
Chebyshev's Rule	The proportion of data within k standard deviations of the mean is never less than $1-\frac{1}{k^2}$, where k is greater than 1. Useful mainly for non-normal distributions; if the data are normal, the Empirical Rule is preferable.
descriptive statistics	The branch of statistics that deals with ways to represent a data set graphically and in terms of center, relative position and dispersion around the mean.
Empirical cdf	Empirical cumulative distribution function: a histogram in which each observation (x value) is paired not with its frequency, but with the cumulative frequencies of all values less than or equal to it.
Empirical Rule	For bell-shaped data, approximately 68% of the observations are within one standard deviation of the mean; approximately 95% are within two standard deviations; and approximately 99% are within three standard deviations. Compare with Chebyshev's Rule.
EX	Expected value of X, computed as a mean with weights applied to the observations.
geometric mean	A measure of centrality, computed as the nth root of the product of n values, used often with annual percent increases.
grade point average	A type of weighted mean in which course grades are weighted by the proportion of credits each course is worth.
harmonic mean	A measure of centrality, computed as n divided by the sum of the reciprocals of n values, used often with distances.

histogram	A graph in which observations (x values) are paired with their frequencies (y values).
inferential statistics	The branch of statistics that deals with ways to use sample statistics to shed light on unknown population parameters, chiefly μ.
interquartile range	$IQR = Q_3 - Q_1$. It is the height of the box in a boxplot. It's also a term in the definition of outliers.
mean	The most common way to define the middle of a list of numbers, defined as $\Sigma x/n$.
median	The middle value of a sorted list of numbers if the size of the list is an odd number, or the mean of the two middle numbers if the size of the list is an even number.
mode	A measure of centrality, defined as the value that occurs most often; it does not always exist.
normal	A type of bell-shaped distribution.
normal distribution table	A table for normal data that matches z scores of observations with left-tail areas.
normality test	A procedure available in many statistical packages such as Minitab that reports whether a given distribution follows a normal distribution.
outlier	An unusually big or unusually small distribution, usually defined in terms of the interquartile range.
percentile	A measure of relative position, a percentile is one of the 99 numbers that divides a sorted dataset into 100 regions; P_{99}, for example, is that value that is bigger than the bottom 99% of observations.
quartile	A measure of relative position, a quartile is one of the three numbers that divides a sorted dataset into four regions. Q_3, for example, is that value that is bigger than the bottom 75% of observations. Note $Q_3=P_{50}$=median.
range	A measure of dispersion, range = maximum - minimum.
skewness	A quantitative measure of departure from symmetry.

standard deviation	The most common measure of spread of data around the mean, defined as square root of average squared deviations from the mean.
uniform	A continuous distribution in which all observations have the same frequency; its distribution curve is shaped like a rectangle.
variance	A measure of dispersion around the mean, defined as the square of the standard deviation.
weighted mean	Same as expected value.
z score	A measure of the distance of any observation from the mean of the distribution, where the distance is measured in standard deviation units.
Σ	Summation operator. Σx means "sum the x values."
σ	Population standard deviation.

Chapter 1 Sample Test #1

1. Which of the following graphs probably represents a normal distribution? There may be none, one, or more than one.

a.

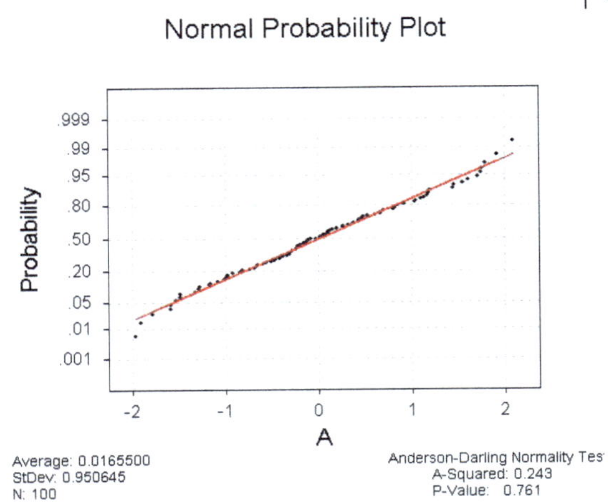

b.

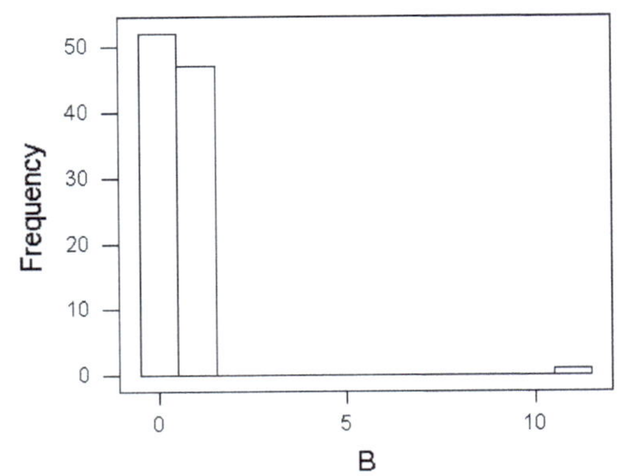

c.

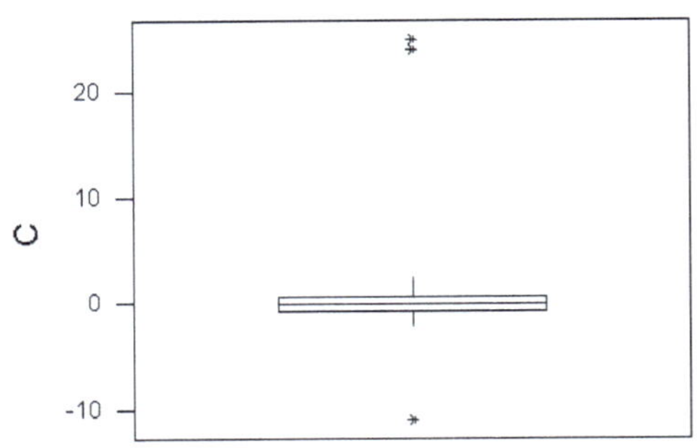

d.

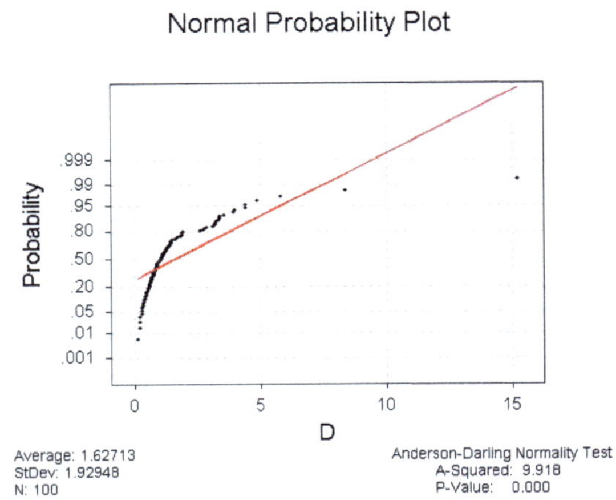

2. Consider the following summary statistics.

Max	110	a. Please report the 5-number summary. b. For the normality test, will $p > 0.05$ or $p < 0.05$?
Median	67.5	
Min	20.5	
P_{90}	105.5	
IQR	15	
Q_1	65	

3.

Please fill in the cells of the table with the correct symbols.		Mean	Standard deviation
	Population		
	Sample		

4.

a. In the following probability table, let X represent the number of master's-level courses taken by a sample of undergraduates, and p(X) is the proportion of undergraduates taking that number of master's-level courses. For example, 20% take two master's-level courses. Please report the $p(X=1)$ and then report the mean, expressed as an integer divided by a power of 10, and the standard deviation.

X	0	1	2
p(X)	0.45		0.20

b. The following table gives the proportion of undergraduates enrolled in various majors. Why is it NOT a probability distribution?

X	Accounting	Nursing	Math	Engineering
p(X)	0.20	0.30	0.40	0.10

Chapter 1 Sample Test #2

1. (10) Consider this report card:

Course	Credits	Grade
English	3	C
Calculus	4	B
History	4	A
Accounting	3	C
Physics	4	C

a. Please compute the student's GPA. Show calculations and answer, rounded to three decimals on your answer sheet.

b. Which would have the greater effect on GPA: if the Calculus grade had been an A or if the Accounting grade had been a B? Justify your answer with calculations.

2. (6) Suppose a student received these marks in a course. The top row indicates total possible points, the number below it the actual mark received. The last box in the second row represents the final exam, which has not yet been taken, so its current mark is x:

T1/50	T2/100	T3/100	T4/150	Pr1/50	Pr2/60	Pr3/75	Final/200
41	89	87	131	40	57	69	x

If the lowest possible A in this course is 90%, what must be the mark on the final exam to earn an A? If an A is not possible, explain why not. Round your answer to three decimals.

3. (18) If the following are never possible for any data set, write NEVER on your answer sheet; if they are always true for any data set, write ALWAYS. If they are true for some data sets but not for others, please create a set of 3-5 integers for which the statement is true.

a. Max < Min	b. P_{90} > Max	c. $P_{50} = Q_2$
d. Q_3+1.5 IQR > Max	e. St. dev. > Variance	f. Mode does not exist
g. Mean does not exist	h. Median does not exist	i. The mean has z score equal to 0

4. (6) Visitors to a certain website arrive according to a Poisson distribution with mean 2/hour. If X represents the number of visitors during a randomly chosen hour, please compute $P(X < 1)$. Round your answer to three decimal digits (e.g., 0.xyz). Hint: $P(X < 1)$ means $P(X=0) + P(X=1)$.

5. (6) Suppose the amount of time people spend typing text messages follows a uniform distribution with max 4 minutes, min 1 minute. If X represents the time of a randomly selected person, find $P(2 < X < 3.5)$. Round your answer to three decimal digits (e.g., 0.xyz). Remember to draw and label an appropriate graph.

7. (15) For each of the following questions, please draw and label an appropriate curve, show your arithmetic, and round your final answer to four decimal digits. All work and answers must be on your answer sheet.

Suppose it is known that the population of 12-week-old kittens is normally distributed with mean weight 13.5 ounces, standard deviation 0.71 ounces. If there are 14,575 kittens in the population, how many of them weigh

a. less than 15 ounces	b. between 10 and 11 ounces	c. more than 20 ounces

8. (3 points each) Each of the following graphs or summary tables refers to a different data set. If the underlying data set is normal, write NORMAL on your answer sheet; if not normal, write NOT NORMAL. If the graph or data are insufficient to make a conclusion about the normality of the underlying data set, write INCONCLUSIVE on your answer sheet. Each data set has 100 observations.

a.

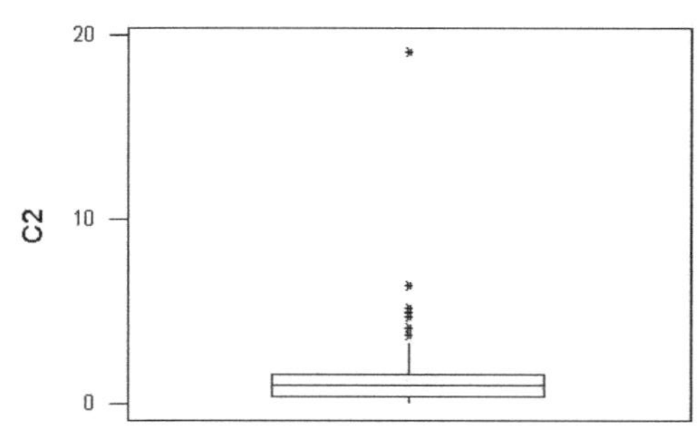

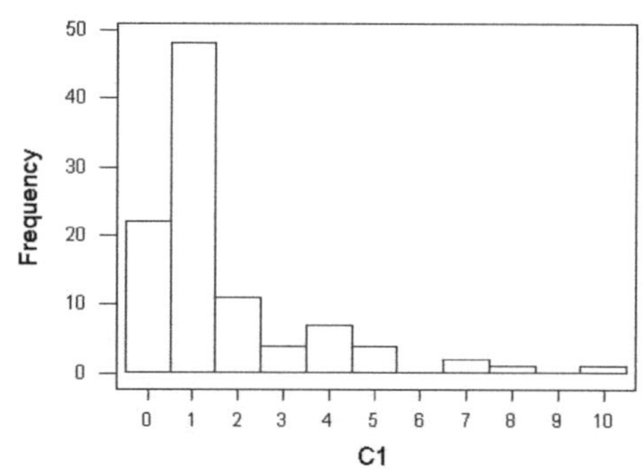

b.

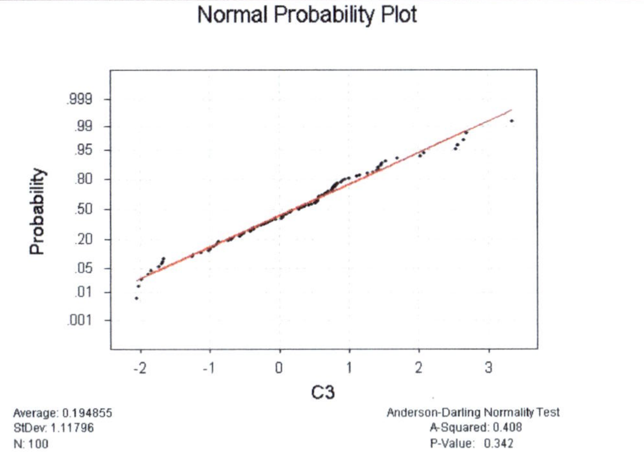

c.

d. Excel reports the skewness coefficient =skew(a1:a100) to be -2.30478.

e. Excel reports the kurtosis coefficient =kurt(a1:a100) to be 6.20808.

f. Max = 10.5835, Q3 = 2.13115, Q2 = 1.01425, Q1 = 0.580525, Min = 0.25.

g. St. dev.=1.118, Mean = 0.195, Max = 3.343, Min = -2.051.

h.

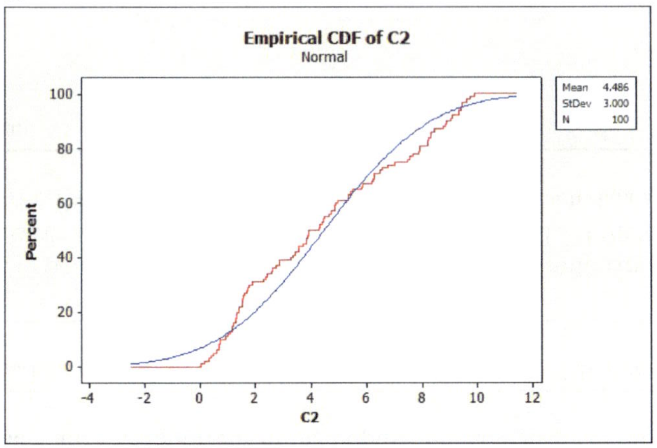

i.

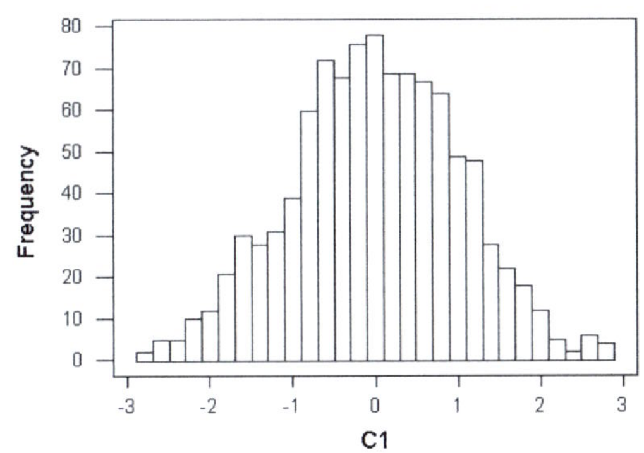

Chapter 1 Sample Test #3

Please write your name on the test sheet, answer sheet, and scratch paper. Hand in all three sheets, but I will only read the answer sheet. Show work and answers on the answer sheet. Round all answers to three decimals or as indicated. Do not use scientific notation. Do not use graphing, programmable, or phone calculators.

1. Compute the grade point average. For a one-grade increase, which course would cause the biggest increase in GPA and why? (10 pts)

Course	Credits	Grade
Linux Programming	2	B
Microeconomics	3	C
German Drama	3	A
Epistemology	3	B
Orchestra	1	C
Abnormal Psychology	4	C

2. Let X stand for the number of visits per week to a local coffeehouse, and p(X) is the proportion of customers who visit the coffeehouse X times per week. For example, 8% of customers visit 7 times per week. Find the missing value of p(X) that will make it a distribution. Please compute the average number of visits, expressed as an integer. Also compute the standard deviation. (10 pts)

X	2	4	5	6	7
p(X)	0.30	?	0.21	0.17	0.08

3. What grade on the final exam is needed for an A in the course? Assume the lowest A is 90%. If an A is not possible, please explain why. (10 pts)

T1/100	T2/110	T3/120	Final/200
82	99	101	x

4. Suppose waiting times at a certain veterinary clinic have $\mu = 6.2$ min, $\sigma = 1.3$ min. (10 pts)

a. Please compute the range of waiting times corresponding to three standard deviations around μ. Express your answer in interval notation.
b. If the distribution is bell-shaped, please report the proportion of data in the interval calculated in part a.
c. Repeat part b if the distribution is not bell-shaped.

5. The times in minutes to complete a pop quiz are summarized below. Please report upper outliers in case there are any. Show all relevant calculations. Would you suppose that the distribution is normal? Would you suppose there is considerable skewness? Justify your answers. (10 pts)

Summary of pop quiz data

Max	21
P_{25}	4.5
P_{75}	9.0
P_{50}	6.5
Min	3.5
P_{90}	19.0
P_{80}	9.5

6. Please select all the true statements about z scores. (10 pts)

a. The mean always has z score equal to 0.
b. Observations greater than μ always have positive z scores.
c. z scores cannot be negative.
d. z scores cannot be greater than 3.5 or less than -3.5.

8. The distribution of gasoline prices is normally distributed with $\mu = 3.32$, $\sigma = 0.14$. Let X represent a randomly chosen gasoline price. Please compute the following probabilities. For each problem, please draw and label an appropriate curve. (10 pts)

a. $p(X > 4.00)$	b. $p(2.00 < X < 4.50)$

9. Choose all the true statements about the following graph. (10 pts)

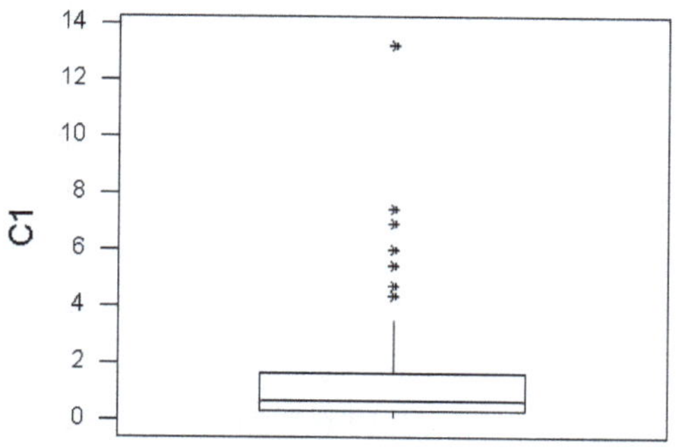

a. The data appear to be non-normal.
b. There appear to be outliers.
c. It appears that mean=median=mode.
d. The data is probably skewed.
e. 68% of the observations will be within one standard deviation of μ.
f. The p value from the normality test will probably be less than 0.05.

10. Choose all the true statements about the following graph. (10 pts)

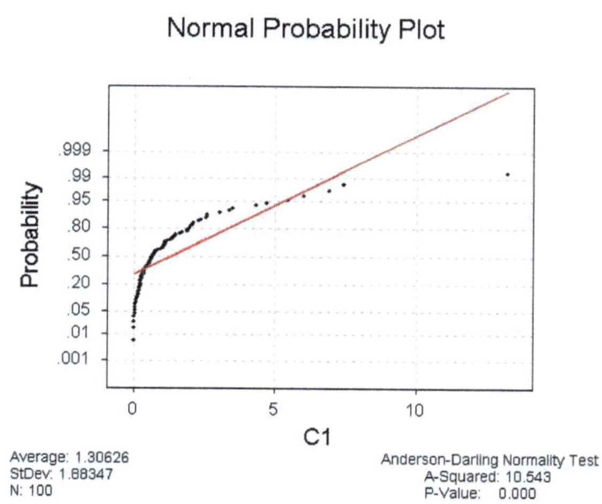

a. There are probably outliers.
b. The z score for the mean will be greater than 0.
c. The max may be less than the min.
d. The Empirical Rule will not apply.
e. The max might be above the upper outlier boundary.

Chapter 1 Sample Test #4

1. Matching (3 points each)

Please write the matching LETTER on your answer sheet. Some numbers might match with more than one letter.

1. {1, 1, 1, 2, 2}	a. range is 0
2. {1, 2, 3, 4, 5}	b. median = mode
3. {99, 2, 1, 2, 4, 2}	c. there is no mode
4. {1, 2, 3, 3, 3}	d. median is 2
5. {4, 4, 4, 4, 4}	e. mode is 1
6. {1, 2, 2, 2, 7}	f. range is three times the median

2. Please answer the following questions about the data set of which this is a summary (4 points each):

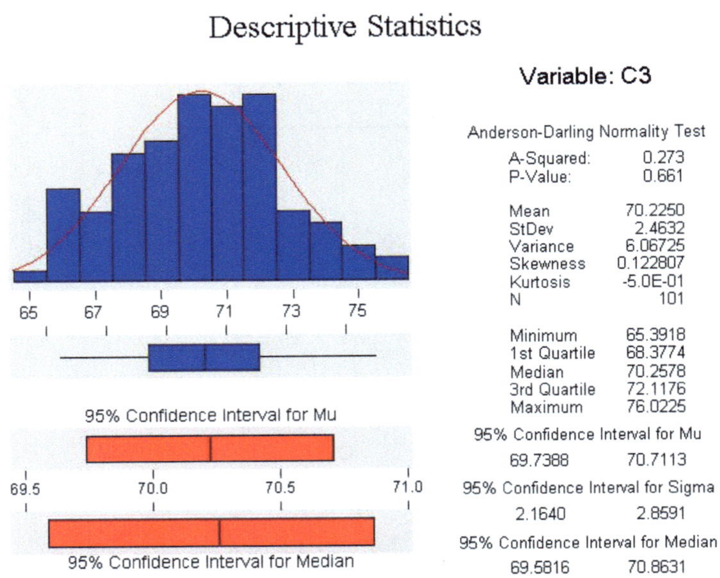

a. T or F: This data is normal.
b. T or F: The max is an outlier.
c. Please compute IQR and round to three decimals.
d. What is P_{50}?
e. What is P_{75}?
f. Assume the max were increased to 999. How would the median be affected?

g. If the max were increased to 999, would the mean increase, decrease, or stay the same?

h. Approximately what is the mode? How do you know?

i. Please compute the range. Round to three decimals.

3. (5 points each) For each problem, please show work and draw and label an appropriate curve. Waiting times in the student cafeteria follow a normal distribution with $\mu = 17.2$ min, $\sigma = 4.1$ min.

a. What proportion of cafeteria customers wait in line fewer than 13.1 min? Please round to four decimals.

b. What proportion wait in line more than 25.4 min? Round to four decimals.

c. What proportion wait between 15 min and 16 min? Round to four decimals.

d. Assume there were 553 cafeteria customers. How many waited in line fewer than 20 minutes? Note that this question is asking for a number of customers, not a proportion.

4. (3 points each) For each of the following statements, write on your answer sheet ALWAYS if it will be true of any data set, NEVER if it is never true (always false) for any data set, and DEPENDS if it is true of some data sets but false for others.

a. The standard deviation can be negative.

b. Max can be less than min.

c. Median does not exist.

d. Mode > median.

e. Upper outlier boundary > Max.

f. Q_2 = Median.

g. Mean is not one of the values in the data set.

6. (5 pts) Let X represent the number of job interviews offered to a particular applicant and p(X) represent the proportion of applicants who receive that number of interviews. For example, 23% of job applicants are offered two job interviews. Please compute the expected value of X.

X	P(X)
1	0.14
2	0.23
3	0.25
4	0.33
5	0.05

Chapter 1 Sample Test #5

1. Compute the grade point average. For a one-grade increase, which course would cause the smallest increase in GPA and why? (5 pts)

Course	Credits	Grade
Linux Programming	2	B
Microeconomics	3	C
Orchestra	1	C
Abnormal Psychology	4	C

2. Let X stand for the number of visits per week to a local coffeehouse, and p(X) is the proportion of customers who visit the coffeehouse X times per week. For example, 8% of customers visit 7 times per week. Find the missing value of p(X) that will make it a distribution. Please compute the average number of visits, expressed as an integer per 100 visits. (5 pts)

X	2	4	5	6	7
p(X)	0.30	?	0.21	0.17	0.08

3. What grade on the final exam is needed for an A in the course? Assume the lowest A is 90%. If an A is not possible, please explain why. (5 pts)

T1/100	T2/150	T3/150	Final/200
80	120	125	x

4. Suppose waiting times at a certain veterinary clinic have $\mu = 6.5$ min, $\sigma = 1.5$ min. (15 pts)

a. Please compute an interval centered around μ corresponding to three standard deviations around μ. Express your answer in interval notation.
b. If the distribution is bell-shaped, please report the *proportion of data* in the interval calculated in part a.
c. Repeat part b if the distribution is *not bell-shaped.*

5. The times in minutes to complete a pop quiz are summarized below. (15 pts)

Summary of pop quiz data:

Max	IQR	P_{75}	P_{50}	Min	P_{90}	P_{80}
21	4.5	9.0	6.5	3.5	19.0	9.5

a. Please report upper outliers in case there are any.
b. Would you suppose that this distribution is normal? Justify your answer.
c. What is the value of the median? Q_2?
d. What number is the upper boundary of the bottom 80% of scores?

6. On your answer sheet, please write the word TRUE or FALSE for each of the following. (8 pts)

a. The mean always has z score equal to 0.
b. Observations greater than μ always have positive z scores.
c. z scores cannot be negative.
d. z scores measure distances from the mean in standard deviation units.

8. The distribution of gasoline prices is normally distributed with $\mu = 3.25$, $\sigma = 0.25$. Let X represent a randomly chosen gasoline price. Please compute the following probabilities. For each problem, please draw and label an appropriate curve. The area to the left of an observation with z score of 1 equals 0.8413; to the left of an observation with z score -2, the area equals 0.0228; to the left of a z score of -1, the area equals 0.1587. (10 pts)

a. $p(X > 3.50)$	b. $p(2.75 < X < 3.00)$

9. Please write the full word TRUE or FALSE on your answer sheet. (10 pts)

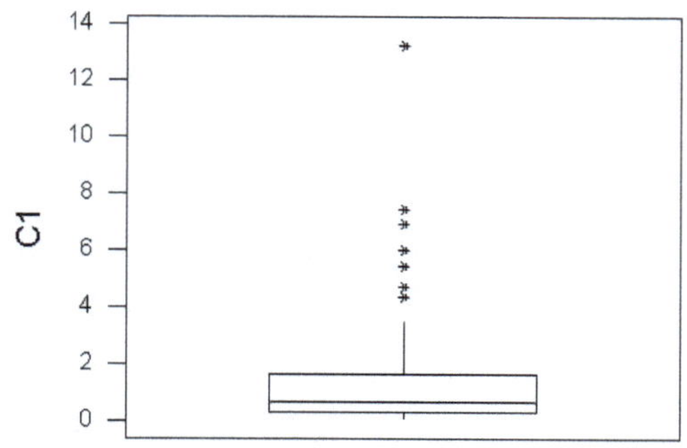

a. The data appear to be non-normal.
b. There appear to be outliers.
c. It appears that mean=median=mode.
d. 68% of the observations will be within one standard deviation of μ.
e. The p value from the normality test will probably be less than 0.05.

10. Please write the full word TRUE or FALSE on your answer sheet. (14 pts)

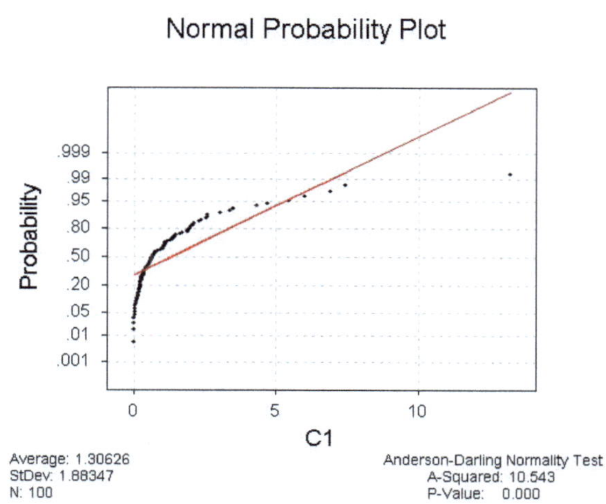

a. There are probably outliers.
b. The z score for the mean will be greater than 0.
c. The max may be less than the min.
d. The Empirical Rule will not apply.
e. Chebyshev's Rule will not apply.
f. The max might be above the upper outlier boundary.

Chapter 2: Confidence Intervals, Hypothesis Testing, and ANOVA

Introduction

Having discussed populations, we turn our attention to the issues and techniques involved with sampling. First, we briefly discuss sampling per se, that is, ways to pick a representative subset from a population.

Next, we will turn our attention to ways of using samples to infer properties of the parent populations. The foundational result, the Central Limit Theorem (CLT), will be discussed, then applied in the estimation of the unknown population mean and later in the testing of hypotheses that the population mean equals a specific value. The difference between estimation and testing is subtle but important: when we estimate a population parameter using an interval, we are describing a range of values that we believe that parameter may actually have; when we perform a test, however, we are claiming the parameter has one particular value and using the sample statistics to inform us whether to reject that claim or not.

The normal distribution is mentioned in the Central Limit Theorem; hence, it plays an important role both in the construction of confidence intervals and in the testing of hypotheses. It turns out, though, that one facet of the normal distribution is problematic: it assumes that the population standard deviation is known. It probably isn't, since the whole point is that the population is unknown. The t distribution is used in situations when the population standard deviation is assumed unknown, and we introduce it here.

An important type of hypothesis test called ANOVA (Analysis of Variance) is suited to experiments in which multiple samples are compared, with the hypothesis that the means of the populations from which the samples were drawn are equal. We will develop the framework for ANOVA problems and present examples.

Sampling

In Chapter 1, we discussed a number of statistical principles and techniques under the assumption that complete population information is available. But this assumption is rarely justified. Almost all the time, populations are far too big and variable to allow direct observation. The next-best alternative is to collect observations on a much smaller random subset of observations that we hope match the population's characteristics. Calculations based on the sample can be used to estimate the characteristics of the population in which we're interested.

Sampling is a deep, rich, and very technical branch of statistics, and in this book we can barely hope to even begin to scratch the surface. There are books, journals, courses, and even entire academic programs devoted to sampling. But to establish the context in which we will discuss inferential statistics, we will need certain terms and concepts from this field.

Sample size is the number of observations in a sample. We will denote the sample size by n. Bigger sample sizes are better, as long as we do not introduce bias. In some sense that we will later make precise, bigger samples contain more information than smaller ones.

A sample is random if the experimenter does not favor any particular observation over any others.

This does not mean that all observations in the population have an equal chance of going into the sample; there may be some probability function that makes certain observations more likely than others. For example, suppose that ¼ of a certain city's population have master's degrees. We would very much hope that the proportion in a random sample would match those of the population with regard to the characteristics involved in our experiment. Thus, we hope that ¼ of the sample would also have master's degrees.

Bias is basically the effect of a sample's characteristics not matching those of the population. Hence, we desire unbiased samples. The problem is that we can't always anticipate where the bias will come from or guard against it. The results from biased sampling can be ghastly. Probably the most famous example of bias happened in the 1936 American presidential election.

Alf Landon, the conservative Kansas governor and Republican challenger, opposed Franklin Roosevelt, the liberal democratic incumbent. The month preceding the election, the *Literary Digest* conducted a telephone poll to predict the outcome. The poll results suggested a 57%–43% landslide victory for Landon, but in the actual voting, President Roosevelt scored a 62%–38% landslide, including all but two states—the biggest landslide up to that point in history, in fact. What happened? The telephone introduced bias, specifically, respondents were selected via the telephone. In 1936, most people did not have telephones, partly because the country was in a major depression. In effect, the poll was heavily biased toward Landon's supporters.

Techniques are vastly improved over those of the 1930s, but it's still worth pausing to consider what might be today's equivalent source of bias. Increasingly, for example, cell phones are supplanting landline phones. A poll designed to contact respondents primarily via landline will probably be biased against the cell phone users.

Self-selection polls are ubiquitous, but beware of them. Many TV shows, web and social media sites, and newspapers invite the public to register their opinions: Who will win the Super Bowl? Will Candidate X win the election? Are taxes too high? The problem is that such results are meaningless unless the questioner selects the respondents. Self-selection polls, in which individuals decide for themselves whether to participate, have no value beyond entertainment.

One more point about election polls: read the fine print. Polls vary by assumptions and technique; there's more than one way to conduct a legitimate, scientific poll. Pay attention to the distinction among the general public, registered voters, and likely voters. Candidate X may outscore Candidate Y among all respondents, but if some of these don't cast a vote, their opinion probably won't affect the outcome. This could be a concern even among registered voters, since not everyone who is registered will vote. Although it's impossible to know exactly who will vote, the best guess is to poll the ones who have voted recently.

Central Limit Theorem

Our main concern is the relationship between sample statistics that can be calculated, hence are known, and the unknown—and unknowable—population parameters. To be more specific, we wish to use the sample statistics to make inferences about the population parameters. The foundational result is the Central Limit Theorem, but first we need to clarify some notation.

Now that we are distinguishing between samples and populations, we would benefit from separate notation for sample means versus population means, and similarly for standard deviations. It's traditional to use Greek letters for populations and Latin for the samples, as follows:

	Mean	Standard Deviation
Population	μ	σ
Sample	$\bar{x}$	s

The symbol $\bar{x}$ for the mean is very common, it should only be used with reference to samples, not to populations. The quantities μ and $\bar{x}$ are computed the same way: add the n observations, then divide the sum by n. The computation of the standard deviation, however, does differ by context: the population standard deviation uses a slightly different formula than the sample version. As we saw in Chapter 1, for populations we compute the standard deviation σ as

$$\sigma = \sqrt{\frac{\sum (x-\mu)^2}{n}}$$

while the sample standard deviation s is computed as

$$s = \sqrt{\frac{\sum (x-\bar{x})^2}{n-1}}$$

The difference between the two has to do with bias, which we discussed above

To resume: we would like to know what can be inferred about μ based on knowledge of $\bar{x}$.

Imagine we collect an infinite number of samples from a population, in such a way that each sample has the same number of observations n. We would compute the mean $\bar{x}$ of each sample, hence we would have an infinite number of such $\bar{x}$s. The CLT describes three properties of this hypothetical, infinite set of sample means:

> **Central Limit Theorem**
> If an infinite number of samples is collected from a population, each of sample size n, and $\bar{x}$ is computed for each sample, then
>
> 1: The infinite set of $\bar{x}$ forms a normal distribution.
> 2: The mean of the normal distribution of $\bar{x}$ is μ.
> 3: The standard deviation of the distribution of $\bar{x}$ is $\frac{\sigma}{\sqrt{n}}$.

If n < 30, use $\frac{s}{\sqrt{n}}$ for the standard deviation of the distribution.

The CLT is the engine running quietly in the background of many important results in inferential statistics, including confidence interval estimation for μ, hypothesis testing for μ, and ANOVA. Because of its immense importance, it is well worth committing to memory. The following interpretive comments may aid understanding.

The CLT is an asymptotic result. It isn't strictly true for any particular value of n, but gets increasingly closer to truth as n increases without bound. This hypothetical result shouldn't bother us, though; other disciplines stipulate idealized, hypothetical results that are only approximately true in reality. In Physics, one derives equations of motion for frictionless surfaces, but frictionless surfaces do not exist in real life. One then progressively complicates the models by accounting for friction and other disturbances. In Economics, one hypothesizes supply-and-demand curves under the assumptions of perfect rationality and complete information and then, as in Physics, introduces complications that make the theories more realistic.

The CLT assumes nothing about the distribution of the underlying population. In particular, it need not be normal.

Students occasionally want to know why we bother estimating μ or hypothesizing about it since point 2 of the CLT assures us that the mean of the sample means equals μ. But of course, point 2 is only approximately true; in light of the previous comment, point 2 (and, for that matter, point 3) is only strictly true in the hypothetical situation in which we actually computed an infinite number of $\bar{x}$s.

Students also are concerned by the presence of σ in the third point. If the population is unobservable, then how can we base a result on the unknowable σ? This is a fair point, albeit the same point we made above: point 3 is true in the limit as n grows without bound but not for any particular value of n. In practice, it turns out that s is a good estimate of σ if the sample size n is at least 30. For smaller sample sizes, we will have to slightly modify our technique.

The CLT emphatically does ***not*** claim that all samples have mean μ. Since we're imagining an infinite number of samples, then there may of course be considerable variation in the distribution of the $\bar{x}$s. Suppose we estimate the population mean income level of Chicagoans based on samples of size n = 100. In the hypothetical framework of the CLT, one such sample may consist, say, of bank presidents. Their $\bar{x}$ would be very impressive. A different sample of size n = 100 might consist of homeless people, or second-graders, or ministers; each of these $\bar{x}$ would be very different from the bank presidents'. The point is that the values of the $\bar{x}$ will vary from sample to sample, but the mean of these $\bar{x}$ is the unknown μ.

The CLT is often abbreviated $\bar{x} \sim N(\mu, \sigma^2/n)$, which is read ***$\bar{x}$ has a normal distribution with mean μ and variance σ^2/n.*** The variance is of course the square of the standard deviation, and ~ means "has the distribution of."

Confidence Intervals for μ

The CLT cannot be used exactly as stated to estimate μ because, of course, we cannot collect an infinite number of samples. If we had the resources to do so, we would have computed μ directly in the first place. What we will actually do is collect a single sample and treat it as though it came from an infinite number of other samples we could have drawn but didn't.

A ***confidence interval for μ*** is a range or interval of values that we will often represent as a segment on the x-axis of the Cartesian coordinate system, and that we believe is likely to contain μ. The confidence interval formula is remarkably short, but before we discuss it, there is a matter to which we must attend: How much confidence shall we attach to the interval we compute? That is, how certain should we be that μ lives in the interval we compute for it? Surprisingly, the answer is not 100%.

To understand why, imagine you have been hired as an educational consultant for some agency in your state. Your charge is to provide an interval estimate of the population grade point average of your state's undergraduates. Colleges vary considerably by mission, admission criteria, grading criteria, and many other factors, so this matter would involve considerable effort would.

Suppose you report to the agency that you have an interval that you are 100% certain contains μ. The members of the agency will initially be delighted, but delight will turn to fury when you report that the interval is [0.00, 4.00]. That is, you are 100% certain that μ is somewhere between 0.00 and 4.00, inclusive. The fury is because your interval only recapitulates the range on which grade point averages are measured in the first place; you haven't told the officials anything they didn't already know.

That's the problem with 100% confidence intervals: the interval range is far too broad to be practical. But there is also a risk in going too far in the other direction. Suppose you attempt to soothe tempers by reporting that you do actually have a serious candidate for an interval: μ is, you claim, somewhere between 2.6248 and 2.6249. Wonderful, the officials tell you, as they offer you a check for your consultant's fee. Just one question: how confident are you that μ is actually in that very narrow interval? Not at all confident, you reply: that interval is so narrow that you'd be very surprised if it contained μ. They told you they wanted an interval narrower than the [0.00, 4.00] you offered them the first time, and you gave them one. At this point, the officials snatch back the check and kick you out of their office.

In confidence interval estimation, we need to recognize the ***trade-off between confidence and precision*** for fixed n. Too much confidence, such as 100% confidence, leads to an unacceptable lack of precision; too much precision, such as the interval of length 0.0001 in the above example, associates with lack of confidence.

The traditional approach to this trade-off is to establish 95% as the default confidence level. Also popular are 90% and 99%, but 95% is the default for most software packages. Ninety-five percent confidence is very high, though we have to live with a small possibility of error; and it gives interval lengths that we can usually live with, though not always as narrow as we might hope.

Note the caveat that the trade-off between confidence and precision applies ***for fixed n***. If we could increase n at will, there would be no trade-off: we could simultaneously achieve high confidence that our intervals would contain μ, and the intervals would be as narrow as we desire. On the other hand, there is no free lunch: increasing n can be difficult and expensive, and we should not take for granted that large sample sizes are practical.

Let's develop the 95% confidence formula in the context of an example.

Examples

Example 1

You've been given one more chance to estimate your state's population GPA. To this end, you survey n = 400 students and compute $\bar{x} = 2.88$, s = 0.62. There are infinitely many other samples of size 400 that you could have drawn; by the first part of the CLT, the $\bar{x}$s of all these samples form a normal curve:

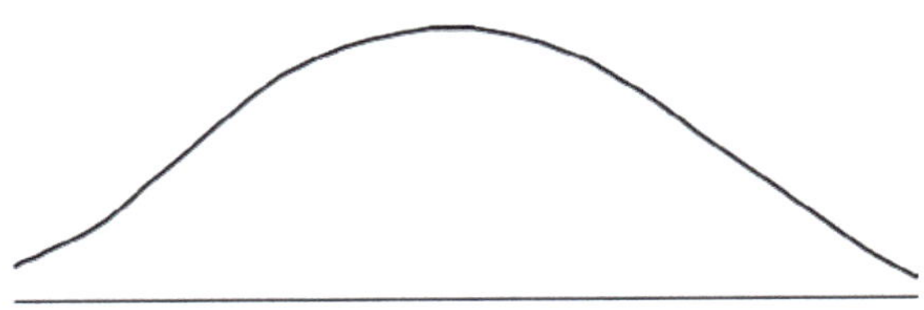

FIG. 2.1

Since we want to be 95% confident that our interval will contain μ, we seek z-score endpoints that bound a middle area of 0.95. Using the symmetry of the normal curve and the fact that the total area beneath the curve equals 1, we see that both tails under the curve have area 0.025. Using a table of normal values as described in Chapter 1, or Excel's normal inverse function, we find that the z-score boundaries for an area centered on the mean are ± 1.96.

What does ± 1.96 represent? Remember that, as points on the x axis, they are z scores; and z scores measure distances from the mean in standard deviation units. Thus, the boundaries of the 95% confidence interval are 1.96 standard deviation units away from the mean on both sides.

At this point in our discussions, the term "standard deviation" could refer to the population σ, the sample s, or, apropos of part 3 of the CLT, $\frac{s}{\sqrt{n}}$. Which one is right?

As a hint, remember what the points on the normal curve represent: means of samples. Every point on the curve described in the first part of the CLT is the mean of a hypothetical sample. Hence, the ± 1.96 z scores are measuring sample standard deviations away from the mean; therefore, we can represent the interval endpoints as

$$(\bar{x} - 1.96\frac{s}{\sqrt{n}}, \bar{x} + 1.96\frac{s}{\sqrt{n}})$$

Returning now to our example, we have $\bar{x}=2.88$, s=0.62, and n=400. Then our 95% confidence interval for μ is

$$(2.88 - 1.96\frac{0.62}{\sqrt{400}}, 2.88 + 1.96\frac{0.62}{\sqrt{400}}) \text{ or } (2.819, 2.941)$$

Our problem is solved: based on our sample of 400 students, our 95% confidence interval for μ is (2.819, 2.941).

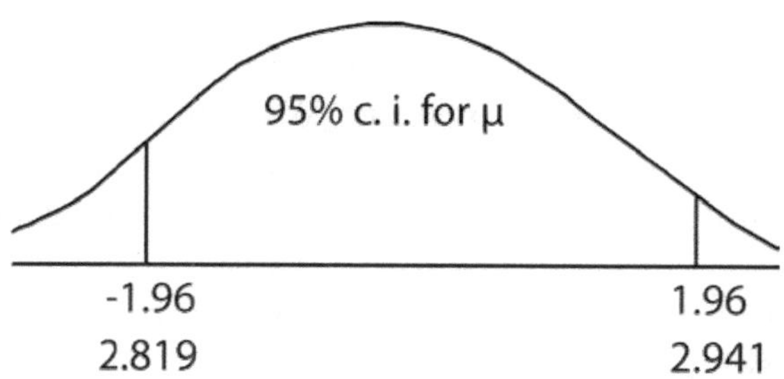

FIG. 2.2

Before considering a second example, let's take a moment to consider some important properties of this problem.

First, please be careful about order of operations. If you're computing it by hand, say for a quiz or exam, first compute $\sqrt{n}$, then divide s by this quantity and multiply the quotient by 1.96. The last step to perform, although it is written first, is to add and subtract this quantity from $\bar{x}$.

Along those lines, it is traditional to express confidence intervals as (smaller, bigger). The term with the subtraction is written first, the term with addition written second; these two are separated by a comma and enclosed in parentheses. Sometimes students compute the "plus" term first and mistakenly write it before the comma.

A bit more substantively, note that this one simple formula $(\bar{x}-1.96\frac{s}{\sqrt{n}},\bar{x}+1.96\frac{s}{\sqrt{n}})$ applies to ***all*** 95% confidence interval problems. The formula does not have to be re-derived from one problem to the next. Also, it applies for all different sample sizes. Whether n is 40, 400, or 4000, the same formula is applied. The importance of this observation will come into focus when we discuss the t distribution formula, which does vary with varying sample size.

The interval we computed for population grade point averages, (2.819, 2.941), has length 0.122. It depends on $\bar{x}$, s, and, in particular, n. We, the experimenters, control the size of the sample. Although we should not take for granted that big samples are easy to obtain, it is instructive to explore the effect of increasing sample size on interval length. Is it obvious that increasing n will increase the left-hand endpoint, shrink the right-hand endpoint, and hence shrink the length of the interval?

If we increased n to, say, 1000, we would compute

$(2.88 - 1.96\frac{0.62}{\sqrt{1000}},2.88+1.96\frac{0.62}{\sqrt{1000}})$ or (2.842, 2.918), which has length 0.076, a 38% decrease.

This sheds light on what I said earlier: there is a trade-off between confidence and precision for fixed n, but not if n is allowed to increase. We can find a sample size n to achieve whatever precision we desire—whatever interval length, no matter how narrow—and still retain 95% confidence, subject to the caveat of costs involved with collecting big samples.

For example, if we wanted the n that would correspond to a 95% confidence interval of length 0.001, we would solve

$$(2.88 + 1.96\frac{0.62}{\sqrt{n}})-(2.88-1.96\frac{0.62}{\sqrt{n}}) = 0.001,$$

$$\frac{2.4304}{\sqrt{n}} = 0.001$$

which gives us n = 5,906,845, rounded. A sample of size six million certainly seems impractical, but the point is that, at least theoretically, we simultaneously achieved high precision and high confidence. Incidentally, this interval would be (2.8795, 2.8805).

A particularly important note about confidence intervals is the role played by the confidence levels. Of what, exactly, are we 95% confident? Surely μ either belongs to the interval or it does not; there is no way for a parameter to be 95% within an interval.

The explanation is that the term "95% confidence interval for μ" is a slight abuse of terminology. The number 95% does not refer to the particular interval we computed, but rather to the proportion of such intervals that would contain μ if we repeatedly resampled and recomputed them. Even worse, there would be no way to know which 95 out of 100 intervals contained it.

The percentage 95% refers to the process of computing multiple intervals, not to any one particular interval. In this sense, the label "05% confidence interval for μ" is misleading. Nevertheless, confidence intervals are universally described this way, and as long as the discrepancy is pointed out, it's not much of a problem.

Example 2

Suppose we wanted to estimate the population mean amount of time spent waiting in the checkout line at your favorite supermarket. If your sample of size n = 113 has $\bar{x} = 4.4$ min, s = 0.9 min, you would compute $(4.4 - 1.96\frac{0.9}{\sqrt{113}}, 4.4 + 1.96\frac{0.9}{\sqrt{113}})$ or (4.234, 4.566). This interval has length 0.332; if we wanted a narrower interval that still has 95% confidence, we could increase the sample size. The unknown population mean μ will belong to 95% of such intervals, but there is no way to know whether or not it is between 4.234 and 4.566. Incidentally, note that we took 95% confidence by default. If a problem does not specify the confidence level, assume it is 95%.

If the confidence level is 95%, then the probability of "making a mistake," in some sense, is 1 – 0.95 = 0.05. A "mistake" could count either as claiming that μ is in the interval when it isn't, or claiming that the interval doesn't contain μ when it does. The two concepts, confidence and "mistake," basically refer to the same underlying phenomenon: to claim 95% confidence is to provide exactly the same information as "5% mistake." Shortly, however—when we introduce hypothesis testing—we will discover a number of advantages in making the change in terminology. In fact, it will be so useful that we will use the term "significance" and give it a Greek letter, α (alpha).

90% and 99% Confidence Intervals for μ

Also commonly used are interval estimates for μ of 90% and 99% confidence. The logic is exactly the same; only the z score boundaries change. For 90% confidence, the tails add to 10%; hence, by symmetry, each individual tail will be 0.05. Using Excel's normal inverse function, we find z scores of ± 1.645, and remembering that we are measuring in standard deviation units—the standard deviation of the curve of means—we obtain $(\bar{x} - 1.645\frac{s}{\sqrt{n}}, \bar{x} + 1.645\frac{s}{\sqrt{n}})$.

For 99% confidence, using the normal inverse function with 0.005 tails, the z scores are reported as ± 2.575, which gives the formula $(\bar{x} - 2.575\frac{s}{\sqrt{n}}, \bar{x} + 2.575\frac{s}{\sqrt{n}})$.

The only term that changes as we change confidence level is the z-score multiplier of $\frac{s}{\sqrt{n}}$. These multipliers are called critical numbers; in the next section, hypothesis testing, we will see what makes them critical.

Common confidence levels and corresponding confidence interval formulas

90%	$(\bar{x} - 1.645\frac{s}{\sqrt{n}}, \bar{x} + 1.645\frac{s}{\sqrt{n}})$
95%	$(\bar{x} - 1.96\frac{s}{\sqrt{n}}, \bar{x} + 1.96\frac{s}{\sqrt{n}})$
99%	$(\bar{x} - 2.575\frac{s}{\sqrt{n}}, \bar{x} + 2.575\frac{s}{\sqrt{n}})$.

Confidence Intervals Using the Graphing Calculator

You can use the TI-83 to report confidence intervals this way:

Choose Stat – Tests – Z interval (option 7). Assuming you have already computed the summary statistics, right-arrow to highlight Stats and enter the values for σ, $\bar{x}$ n and the confidence level. Then choose Calculate.

Hypothesis Tests for μ

An alternative approach to confidence interval estimation is to state a claim that μ equals a particular value, then use the sample data to assess. The claim is called the ***null hypothesis***, H_0. Although computationally very simple, there are many important interpretive details.

In what follows, we will need to introduce the concept of significance. The significance of a test is 1 - confidence. For instance, 95% confidence is the same as 5% significance, or α = 0.05. Let's illustrate with an example.

Suppose it is claimed that the population mean time to complete a stats final exam is 70 minutes. Based on a sample of n = 198, we obtain $\bar{x}$ = 72.5, s = 9.2.

The basic idea is that if $\bar{x}$ is close to the hypothesized μ, the support for H_0 will be strong, otherwise not. We will require the z score for the sample mean because, of course, the unit of measurement is the standard deviation $\frac{s}{\sqrt{n}}$: we are still operating in the context of the CLT in which the data are means of samples that are normally distributed with mean μ and standard deviation $\frac{s}{\sqrt{n}}$. By z score, we mean $z_{\bar{x}}$, which in this case is $z_{72.5}$.

Thus, the difference between $\bar{x}$ and μ is $z_{72.5} = \frac{72.5-70}{\frac{9.2}{\sqrt{198}}} = 3.82$. But we need a basis of comparison, a threshold, to determine whether it is big or small. For this, we will call on 1.96, the critical value for 95% confidence. (Remember that 95% is the default confidence level.) The numbers ± 1.96 were the boundaries of a region of 95% confidence, so it is at least reasonable to extend this reasoning to say that any $z_{\bar{x}}$ that exceeds 1.96 in absolute value represents an $\bar{x}$ that is too big for the confidence interval, hence opposes H_0. (We refer to absolute value because we care only about the magnitude, not the direction, of the difference. If $\bar{x} > \mu$, then $z_{\bar{x}}$ will be positive, otherwise negative; it is to avoid duplicating such conditions that we state it with absolute value.)

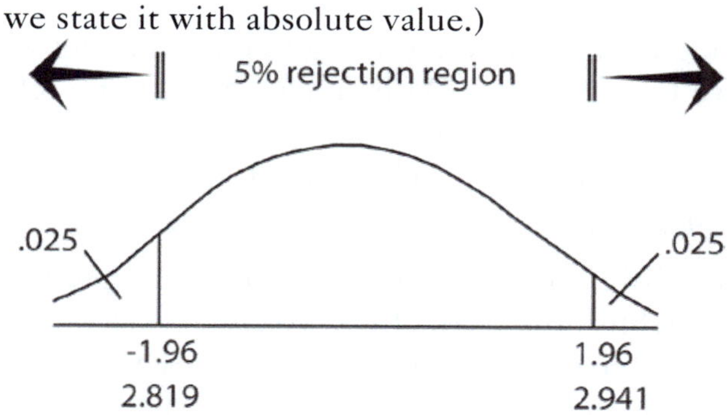

FIG. 2.3

The Alternative Hypothesis H_1

We also should stipulate what it is that we will be claiming about μ in case we reject H_0, which we will call the alternative hypothesis H_1. This may initially seem subtle but can have a major impact on our conclusions. To digress for a moment, suppose there is a study of heights of undergraduates, and for this study, we have H_0: $\mu = 60$ inches. Although there may well be individual students who are five feet tall, it would be very odd to claim that this is the population mean; but no mathematical principles would be violated, so we proceed with our example.

If the sample data leads us to reject the null, what exactly would we be saying about μ? That it is less than five feet? Greater? Split the difference between < and >? Knowing what we do about typical heights, we would probably choose greater, so H_1: $\mu > 60$. If it is false that the population mean equals five feet, surely it will be the case that it is greater.

On the other hand, suppose it is claimed that the population mean age of undergraduates is 30 years: H_0: $\mu = 30$. This would also be very odd since, although there certainly are undergraduates who are this age, it's hard to believe that μ is this big. If the null is false, then what are we saying about μ? Surely the alternative should be that μ is less than 30: H_1: $\mu < 30$.

Sometimes we can't make up our minds about whether to use < or > for H_1, so we literally split the difference and state the alternative with ≠. This time, suppose we hypothesize that the population mean weight of an undergraduate student is 170 lbs.:H_0: $\mu = 170$. It doesn't seem unusually high or unusually low, so if H_0 is false, we will have to split the difference: H_1: $\mu \neq 170$. If it is false that the population mean is 170, then it is as likely to be higher as it is to be lower.

To reiterate, the algebraic sign we use in stating H_1 can have a major impact on our conclusions. For the moment, we will state our alternatives with ≠ since we want to preserve the connection with our discussion of confidence intervals. But before we finish, we will revisit this discussion and study the effect of the so-called one-sided, or one-tailed, alternative hypotheses.

Rejecting versus Failing to Reject

We'll need one more digression before returning to our example: we almost always state null hypotheses that we hope to reject. Without cheating—purposely looking for samples that support our biases or stating ridiculous null hypotheses,—we hope that a representative sample will cause us to conclude that H_0 should be rejected: in other words, that $|z_{\bar{x}}|$ will exceed 1.96, assuming 5% significance and two-tailed alternative hypothesis.

On the other hand, we should never state that we accepted a null hypothesis. The sample data can never establish the truth of H_0: we either reject it or fail to reject it. Most textbooks compare this paradigm to the legal system: juries either find that defendants are guilty or they fail to do so, but it is extremely rare to conclude that a defendant was innocent of charges. Hypothesis testing works the same way.

The conclusion to a hypothesis test is either to reject H_0 or fail to reject it. The conclusion is not the value of the test statistic, although of course it is essential that we know this value. The conclusion is also not a statement about the alternative hypothesis H_1. To perform any hypothesis test, we compare the test statistic, in this case $z_{\bar{x}}$, with the critical value, in this case ± 1.96. If the absolute value of the test statistic is the bigger of the two, we reject H_0; if the critical number of the bigger of the two, we fail to reject it.

Back to the example.

We tested H_0: $\mu = 70$ against the two-sided alternative H_1: $\mu \neq 70$ based on sample data $\bar{x} = 72.5$, $s = 9.2$, $n = 198$. The test statistic is $z_{72.5} = 3.82$. Since $3.82 > 1.96$, we reject H_0, which is as we hoped.

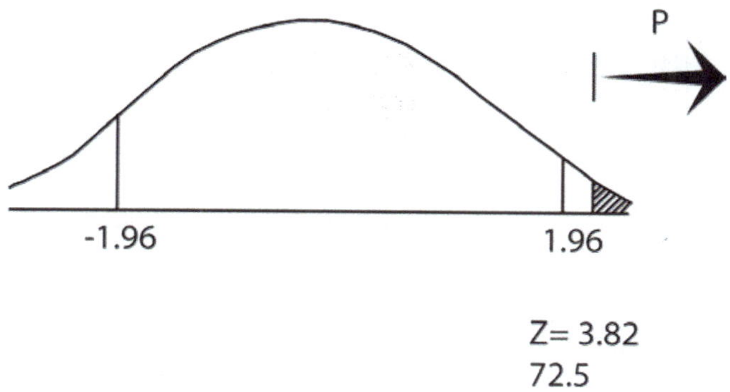

FIG. 2.4

Hypothesis Tests Using the Graphing Calculator

Have your summary statistics available, then start with Stat – Tests. Option 1 will allow you to do the z test—we will perform the t test later. Choose option 1, then with Stats highlighted, enter your values for μ, $\bar{x}$ and n. You will be given the option to select the algebraic sign for H_1 (≠, < or >), then choose Calculate.

Type I and II Errors, Significance Levels, p Values

There remains a small possibility that our conclusion is erroneous. That is because we have chosen not to perform our tests at 0% significance. (Remember significance = 1 – confidence, so 0% significance = 100% confidence.) If our conclusion is erroneous, then we would have mistakenly rejected H_0; that is, H_0 would have been true, but based on our sample statistics, we would have rejected it. Whenever we reject a true H_0, we make an error that has traditionally been labeled an error of Type I.

Remember: one would never make a Type I error if we knew H_0 were true. One never makes an error, on purpose. The most that can be said is that ***if*** our conclusion is erroneous, ***then*** the error of rejecting H_0 would be of Type I.

If we fail to reject H_0, then we are in danger of making the opposite error: an error of omission. If H_0 really is false, but based on sample statistics we did not reject when we should have, we have made what is called an error of Type II—to mistakenly fail to reject H_0. As with Type I errors, we do not know whether or not we ever actually make a Type II error; the most we can say is that in a hypothesis test for which our conclusion is to fail to reject H_o, we run the risk of failing to reject when we should have.

Sometimes I ask my students on tests or assignments to complete the cells in this table:

	Conclusion: Reject H_o	**Conclusion: Fail to reject H_0**
H_0 is true	Type I	no error

H_0 is false	no error	Type II

In a beginning statistics course, there is little more to do with Type I and Type II errors than to define these terms and report them as possibilities when we perform our tests. But in more advanced work, certain statistical tests are often created subject to minimizing either or both error types, which has a quantitative effect on calculations or formulas.

There are various equivalent ways to describe a Type I or Type II error. Assume the null hypothesis is the claim $\mu = k$.

Synonyms for Type I error	Mistakenly reject H_0
	Reject H_0 when it is true
	Reject a true null hypothesis
	Reject H_o when μ really does equal k
	Claim that $\mu \neq$ k when it actually does
Synonyms for Type II error	Mistakenly fail to reject H_0
	Fail to reject H_0 when it is false
	Fail to reject a false null hypothesis
	Fail to reject H_0 when μ does not equal k
	Allow the possibility that μ = k when it does not

Recall that because of the trade-off between confidence and precision for fixed n, we established the default 95% confidence level. We made the side-comment that in hypothesis testing, we represent this concept by defining the term significance **α= 1 − confidence**. Then 95% confidence corresponds to 5% significance; both are considered defaults, by which I mean that if no confidence level has been stated in a hypothesis test, we will take it to be 5%.

The establishment of a significance level, in combination with the shape of H_1 (single-tailed or two-tailed), determines the so-called critical numbers. At the moment, the critical numbers are ± 1.96, although they often change, as we will see. The importance of critical numbers is that they determine the threshold for rejecting H_0; in other words, they determine whether the difference between $\bar{x}$ and μ is "big." For any hypothesis test, we reject the null hypothesis if the test statistic is bigger in absolute value than the critical number, and we fail to reject H_0 if the critical number is the bigger of the two. This function is what makes critical numbers "critical."

There is another approach to hypothesis testing that does not rely on significance levels or the resultant critical numbers. Instead, report the probability of obtaining a result that is even more extreme, or less likely, than the outcome you actually obtained. It is a way of indirectly quantifying the probability of the event in question: if the probability of getting a more extreme event is very small, then the event in question must itself have been unlikely; if the probability of getting a more extreme event is fairly small, then the event in question must have been fairly common. The probability of getting a less likely event is called its **p value.**

Suppose you have a friend who is so tall that the probability is 0.001 of finding someone taller. That must mean that your friend must be unusually tall to begin with, and 0.001 is his/her p value.

Or suppose you have a friend so intelligent that the probability is 0.35 of finding a more intelligent person. Then it isn't hard to find such a person, and your friend's intelligence is mediocre. His/her p value is 0.35.

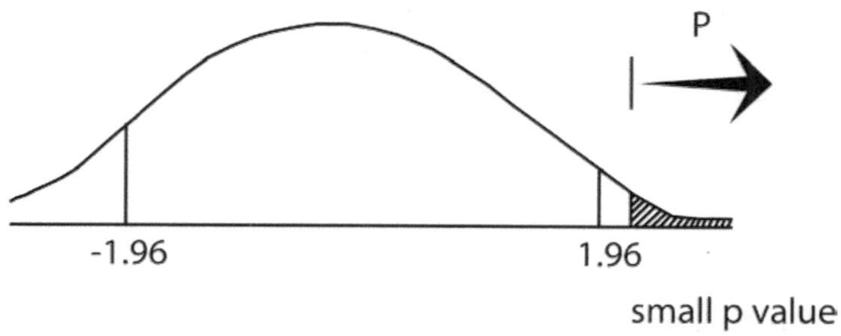

FIG. 2.5

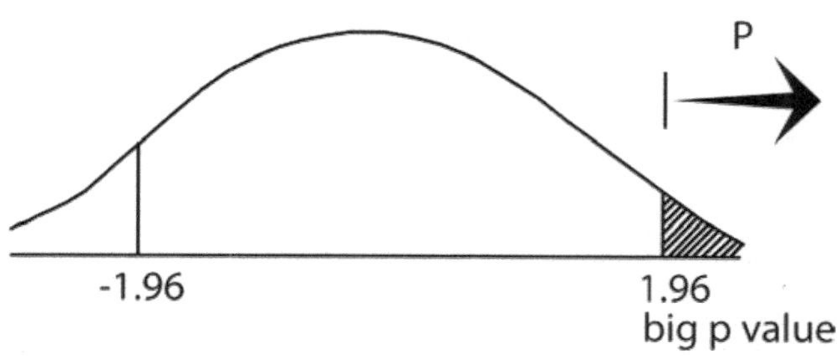

FIG. 2.6

The point is that with p values, we don't need to stipulate an alpha level or find a critical value. Further, the experimenter can determine the p value that associates with rejecting the null hypothesis. One person may reject if $p < 0.05$, someone else may choose $p < 0.02$, and a third person may reject if $p < 0.09$.

Technically, we should say that, for two-tailed H_1, the area corresponding to the p value is twice the area beyond the test statistic. In other words, two tails comprise the p value. The basic concept of "area beyond the test statistic" remains, however, and for single-tailed tests the p value indeed does have a single tail.

The critical value approach and the p value approach are independent paradigms for hypothesis testing. Either may be undertaken without reference to the other. We can make them conform as follows:

	p value	Critical value
Reject H_0	$p < \alpha$	test statistic > critical value
Fail to reject H_0	$p > \alpha$	test statistic < critical value

Most software will report p values for us; they're much too hard, most of the time, to compute by pencil and paper. I do ask my students to shade the p-region beneath the distribution curve and/or to report whether or not $p < \alpha$. In courses in mathematical statistics, however, in which one is learning the formulas and properties of the distribution curves, it can be instructive to use methods of advanced calculus to explicitly compute p values.

It shouldn't surprise us that statistics has two paradigms for hypothesis testing. Lots of fields have more than one paradigm. In economics, one might believe in free markets, or feudalism, or one might be a capitalist or a Marxist; in philosophy, one might be an existentialist, a metaphysicist, or an atheist.

Example 3

We wish to gain some insight on the population mean amount of fines at your local public library. A sample of 102 patrons reveals $\bar{x} = 5.56$, $s = 0.70$. Let us test H_0: $\mu = 5.00$ against the two-sided alternative H_1: $\mu \neq 5.00$.

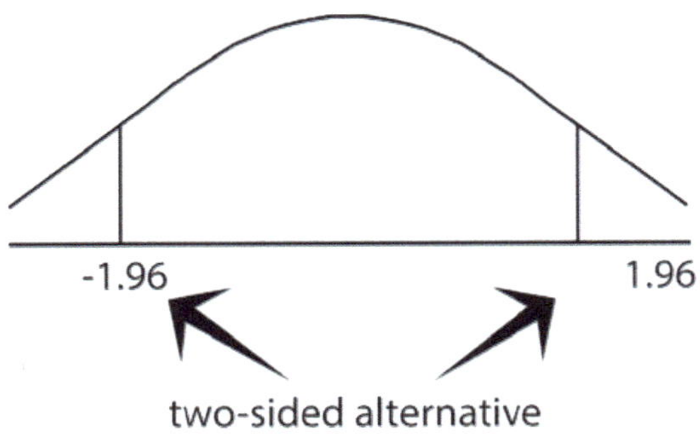

FIG. 2.7

$$z_{5.56} = \frac{5.56 - 5.00}{\frac{0.70}{\sqrt{102}}} = 8.08$$ That means the sample mean 5.56 is 8.08 standard deviations above the hypothesized population mean 5.00, where we are using $\frac{0.70}{\sqrt{102}}$ as the standard deviation (the third part of the Central Limit Theorem). Because the z score is bigger than the critical number—much bigger, in this case—we reject H_0. There is a possible Type I error, and the p value region has been shaded.

Other Significance Levels and One-Tailed Alternative Hypotheses

This procedure is easily adapted to the cases of 10% and 1% significance. Remember that the significance level determines the critical number but does not affect the test statistic, so if we were to change from $\alpha = 5\%$ to $\alpha = 10\%$, all we'd be doing is changing the critical number from 1.96 to 1.645; if we change to $\alpha = 1\%$, the new critical number would be 2.575.

That isn't to say that the conclusion could not change by changing α, only that the calculation—the test statistic—would not change. Suppose we perform a test with two-tailed alternative at $\alpha = 5\%$ and compute a test statistic of 2.01. We would reject H_0 because the test statistic exceeds the critical number. But if we changed α to 1%, changing nothing else about the problem, our conclusion would be to fail to reject because the new critical number would be 2.575, which is greater than the test statistic. Conclusions to hypothesis tests depend on the comparison between the test statistic and the critical number, and α determines the critical number. In other words, α determines what we count as "big."

Some test statistics are so big that they would cause us to reject H_0 for any critical number. If the z score were, say, 11, then any reasonable α would produce a smaller critical number. Similarly for small test statistics: if $z_{\bar{x}}$ were, say, 0.45, that would mean the observation were so close to the hypothesized μ that no reasonable value of α would lead us to reject the null.

Whether or not to reject H_0 also depends on the form of H_1. Remember that we began our discussion by assuming two-tailed alternatives, with the area defined by the significance level evenly divided. We did it this way because we had started by constructing intervals that had left and right endpoints, but as previous examples demonstrated, two-sided alternatives don't always make the most sense.

We may well want the alternative to have all the area in one or the other of the tails, instead of both, based on our knowledge of the data. Consider the example of heights of undergraduates: if it is false that the population mean height is five feet, then surely it would be the case that μ is bigger than five feet. But by changing the H_1 from ≠ to >, we must be changing the critical number, too: when H_1 was stated with ≠, the 0.05 area was divided into halves, giving each half area 0.025. The corresponding critical number, 1.96, is the boundary of such a 0.025 tail. But if all the 0.05 area is in the right-hand tail (the direction of the tail always follows the inequality arrow in H_1), then the boundary must get "pushed to the left," relative to the 0.025 tail. Convince yourself, by drawing curves and using the Excel commands, that the critical number for a right-tailed H_1 with 5% significance is 1.645. By symmetry, the critical number for a left-tailed H_1 with 5% significance is -1.645.

But just as changing α could affect the conclusion, so could changing the alternative hypothesis from two tails to one. If the test statistic were 1.80, we would have failed to reject with two-tailed H_1 since $1.80 < 1.96$ but would reject if H_1 were stated with > because the critical number would decrease to 1.645.

The t Distribution

Let's recapitulate: we started our section on inferential procedures by stating the Central Limit Theorem, according to which, under certain assumptions, sample means follow a normal distribution with mean μ and standard deviation $\frac{\sigma}{\sqrt{n}}$. We used this theorem to derive formulas for the interval estimation of μ for 95% confidence, then for 90% and 99%. By inverting the confidence interval procedure, we developed a testing procedure to assess claims of the form μ = k, where k is some constant. Among the many properties of hypothesis tests are the facts that our conclusions of Reject or Fail to Reject can be changed by changing the significance level, or by changing the alternative hypothesis from two-sided to one-sided. Now we add another wrinkle.

The normal distribution assumes that the population standard deviation σ is known. This is a little too optimistic, especially considering that we have been using the sample standard deviation s instead. It is true that s is a good estimator of σ when n > 30, but most statisticians would prefer to err on the side of caution and modify the procedure to account for the fact that σ is being estimated, not measured. That is, they would like a way to test hypotheses more conservatively. By this I mean that stronger evidence should be required for rejecting the null hypothesis; and by "stronger evidence" I mean bigger critical values. This is the reason for the ubiquitous use of the t distribution.

Like the normal, the t distribution is bell-shaped, symmetric, and has area equal to 1, but the t has fatter tails. The effect is that for any given sample size, critical values from the t distribution will be greater than the corresponding normal ones. Test statistics would have to achieve a greater threshold for rejecting the null.

Excel has a function that reports t critical numbers: = tinv(***probability***, df). "Probability" is the alpha level, and df stands for degrees of freedom. Unlike the normal, the t distribution associates a different critical number for each sample size; recall that the normal uses 1.96 (assuming 5% significance and two-sided H_1) for all sample sizes. For the t, we compute df as one less than the sample size: df = n – 1. In the limit as n grows without bound, however, the t critical values do approach the normal. This is one example of the normal distribution acting as a limiting case of another, more complex one; we will again use the normal as an approximating distribution when we study the binomial distribution in Chapter 4.

To test a hypothesis using the t distribution, follow the same steps as before (although the test statistic should be called t scores, not z scores): compute the difference between the sample mean and the hypothesized μ, then compare with the t critical number. If the t score is the bigger of the two, reject H_0.

The 2-sample t

The t test can be used to compare means of two samples, either paired or unpaired. Samples are said to be paired if there is a natural correspondence. For example, a diagnostic test might be given at the beginning and end of a math course. Each individual would then have two scores, with the second ideally being better than the first.

To perform the 2 sample t in this case, we would create a column of differences for each individual, then use the t distribution to determine if this set of differences is significantly different from 0. In the following screenshot, Excel returns the p value of 0.000012 which, since it is less than α, tells us to reject the null that the difference is zero. This is good: We certainly hope that there has been a significant increase in achievement!

Results from diagnostic test			
100 points possible			
	Pre-	Post-	Diff
1	45	80	35
2	30	79	49
3	52	92	40
4	69	92	23
5	51	81	30
6	12	78	66
7	82	89	7
8	78	84	6
9	59	69	10
10	50	74	24
11	40	98	58
12	83	85	2
13	19	80	61
14	31	61	30
15	30	92	62
16	42	84	42
2 Sample t			
H0: delta = 0			
1.20068E-05			

After entering the values for the DIFF column, we would use the t.test() function to return the p value. The syntax is (t.test(array 1, array 2, # tails (1 or 2), paired=1).

On the graphing calculator, we would choose Stat-Tests-2 Sample t (option 4) and enter our summary statistics for samples 1 and 2.

Power of an Hypothesis Test

We've had much to say regarding hypothesis tests, test statistics, critical values, p values, and Type I and II errors. Of course, Type I and Type II errors are to be avoided, if possible.

What remains to be discussed, and is probably the most important topic in this unit, is the probability of doing what we wanted: correctly reject H_0, that is, correctly to reject false claims. That is called the ***power*** of a test.

Suppose we initially claim H_0: $\mu = k$. Then we can use the sample information to construct, say, a 95% confidence interval for μ. But imagine that we then change our minds and claim instead that $\mu = k_1$. The initial rejection region, which was symmetric under H_0, becomes distorted from the perspective of the new, correct hypothesis. It is what the rejection region looks like when we correctly identify the null hypothesis—in other words, the probability of correctly rejecting the null.

Computing the power in this simple case is straightforward: compute the endpoints of the rejection region from the initial null hypothesis. When the curve is shifted to account for the new value of μ, compute the distances (z scores or t scores) and then use Excel to compute the left- and right-tail areas. The sum of these areas is the power of the test.

To be precise, the sum of the areas is the power of the test H_o: $\mu = k$ against the claim $\mu = k_1$. The power is different for each different value of the alternative. One can draw a power curve, a graph that associates a different value of the power for each different alternative.

Computing the power can be amazingly difficult for complex designs, as with ANOVA. Applications like Minitab have options for computing power, so I would advise the reader to refer to the software user manuals or support sites.

One-Way ANOVA

Suppose there are four coffee shop branches, all under the same ownership—that is, they are franchised. They are intended to give the exact same quality of product and service. Management decides to experiment: it will record the service time of a random sample of six customers at each of the four branches to determine whether the branches are indeed taking the same amount of time to serve their patrons or if, conversely, at least one of the shops is too slow (or fast).

The hypothetical data for this experiment are recorded in this table:

A	B	C	D
4.5	5	9	2
3	5	11	5
3	4	9	6
5	4	11	7.5
6	5.5	12	6
4.5	4	11	5

What should be immediately obvious is that the four shops are indeed not taking the same amounts of time to serve their patrons, but let's be more precise. A small amount of variation in service times is in line with our expectations; indeed, if the same store, with the same baristas and the same customer orders, were repeatedly sampled, the times would still vary. What we're looking for is not whether the times are exactly the same, but whether they are sufficiently different to suggest some systematic cause: a poorly trained barista, a lack of supplies, etc. We are interested not in whether there are differences—there always will be—but in whether the differences are ***significant*** Once again set up a hypothesis test. But this one has a twist.

What we wish to test is not whether the population mean μ equals a particular numeric value, as in our previous hypothesis tests, but whether the population means from the four samples are equal to each other: $H_0 : \mu_A = \mu_B = \mu_C = \mu_D$. Note that no particular value has been specified. This procedure tests for the equality of means, not their value(s).

Incidentally, all our examples will have same-sized samples. This is not required by the theory or by Excel; different-sized samples are fine, with three caveats: First, once we finish discussing ANOVA, we will encounter a technique that does require same-sized samples. Second, Two-Way ANOVA, which follows, does indeed require same-sized samples. Third, since the experimenter is in charge of the sample sizes, (s)he usually finds it easier to make them all the same to avoid any complications later on.

In case H_0 is rejected based on the evidence, we must stipulate what it is we are saying about the μs. There are several ways to do this, but we will follow convention and stipulate H_1: at least one μ differs from the others. That is, if it's false that all the population means are equal, we will say that at least one of them is significantly greater or smaller. This alternative hypothesis is actually a composite because it contains many cases. Maybe one μ differs, but maybe two of them do, or three, or maybe all four have statistically significant differences. Many beginning statistics students stumble on this point: they think that the alternative hypothesis is the claim that all the means are different H_1: $\mu_A \neq \mu_B \neq \mu_C \neq \mu_D$. But this is only one possibility among many.

It's also important to point out that we are testing for the equality of population means, not sample means. The sample means are always observable—and different—hence there is no need to hypothesize about them. What we want to know, but are unable to directly observe, is whether the means of the populations from which they are drawn are equal or not. Sample means can be slightly different but still come from populations with equal μs.

Of course, any hypothesis test must be conducted with observable data, which in this case means the $\bar{x}$ s:

4.333333	4.583333	10.5	5.25

As predicted, all four samples have different $\bar{x}$ s; we predict sample C, with mean 10.5, will cause us to reject the null. But let's not get ahead of ourselves.

Let's think about a way to conduct such an experiment. Imagine computing an estimate of the variance in two ways: first, by using the means of the different samples; second, by considering the entire set of 24 observations as one sample. If grouping into samples A, B, C, and D did not isolate unusually big or small measurements, then the two variance estimates should be roughly equal. That would represent a poorly chosen experimental design; after all, the point of this experiment was that we suspected at least one of the coffee shops had service times out of line with the others, and we collected data to find out which one. But if the service times are basically indistinguishable from each other and from the entire data set of service times, then we will not have the evidence to support the conclusion we suspect.

A variance estimate, throughout all of statistics, is a sum of squares divided by some integer that's generically referred to as "degrees of freedom." To compute the variance estimate based on the means of the samples—the between variance estimate—we shall define both a sum of squares for the between and the degrees of freedom for the between:

$SS_{between}$ = each sample mean minus the grand mean, squared, multiplied by the size of the sample

$df_{between}$ = number of samples - 1

variance estimate = $MS_{between} = SS_{between} / df_{between}$

Note that Mean Square, or MS, is a synonym for variance.

Let's compute $MS_{between}$ for the above experiment:

The grand mean, or the overall mean of all 24 observations, is 6.17, rounded.

$SS_{between} = 6(4.33 - 6.17)^2 + 6(4.58 - 6.17)^2 + 6(10.5 - 6.17)^2 + 6(5.25 - 6.17)^2 = 152.9$, rounded

$df_{between} = 4 - 1 = 3$

Therefore, $MS_{between} = SS_{between} / df_{between} = 152.9 / 3 = 51$, rounded.

Computing the variance estimate based on all 24 observations, which we will call the within estimate, we again will need an SS and a df. SS_{within} is the sum of the squared differences of each observation from its sample mean. This is easy to conceive but complicated to carry out; there will have to be a term for each observation in the experiment, and that can become tedious. The within degrees of freedom, df_{within}, is the total number of observations minus 1, which in this case equals 23. The within variance estimate, MS_{within}, is again the ratio of the corresponding SS to its df.

The table below shows the squared difference between each observation and its sample mean:

0.027778	0.173611	2.25	10.5625
1.777778	0.173611	0.25	0.0625
1.777778	0.340278	2.25	0.5625
0.444444	0.340278	0.25	5.0625
2.777778	0.840278	2.25	0.5625
0.027778	0.340278	0.25	0.0625

SS_{within} is the sum of these decimals, or 33.4, roughly.

$df_{within} = (6\text{-}1) + (6\text{-}1) + (6\text{-}1) + (6\text{-}1) = 20$

Therefore, the within variance estimate, $MS_{within} = 33.4 / 20 = 1.67$.

If we had computed roughly the same variance estimate each way, between or within, we would have evidence to support the null hypothesis—which we definitely do ***not*** want to do. We want to reject it! Our two mean squares are roughly 51 and 1.67. They certainly do not seem close, but we need an objective way to decide.

When two numbers are roughly equal, their ratio will be close to 1; if we were clever about separating out the big (or small) observations from the others in our choice of sample categories, then $MS_{between}$ would be greater than MS_{within}; thus, their ratio would be much greater than 1, or at least greater than some established "definer of bigness." In other words, let us compute the ratio of the two variance estimates, then compare with—you get it—a critical number that defines "bigness."

$MS_{between} / MS_{within} = 51 / 1.67 = 30.5$, rounded.

Defining the critical number is a little involved; there are widely available distribution tables. Excel makes it particularly convenient: the function is =finv(α, $df_{between}$, df_{within}). For our problem, Excel reports the critical number as roughly 3.1. The symbol α refers, as always, to our significance level, and significance = 1 – confidence. By default, we take $\alpha = 0.05$

Let's pause to review what we've accomplished:

It's been hypothesized that at least one coffee shop has a population mean service time that differs from the others. To that end, we imagine collecting a sample of service times from each location—the facts that there are four samples and six observations per sample are not essential—and compute the variance estimates, or mean squares, for the between (based on sample means) and the within (based on each individual observation). If the ratio $MS_{between} / MS_{within}$ exceeds the critical number for this test, which we obtained from Excel, then we would reject the claim of equal population means. It does, so we do reject, and the problem is finished. Or nearly so.

The name of the procedure associated with this hypothesis test is called One-Way Analysis of Variance, or ANOVA. Some people call it Single-Factor ANOVA. It typically appears much closer to the end of most introductory statistics books, but I include it here to stress its relationship with other hypothesis tests: they all involve the stating of null and alternative hypothesis, the calculation of a test statistic that's based on the data, and a threshold or critical number that comes from a distribution table or software package. If the test statistic is the greater of the two, the null hypothesis of equality (or, generically, "no effect") is rejected, otherwise not. (Remember that we typically design experiments and state null hypotheses that we intend to reject.)

It's useful to summarize our analysis in an **ANOVA table**:

ANOVA						
Source of Variation	*SS*	*df*	*MS*	*F*	*P-value*	*F crit*
Between Groups	152.9167	3	50.97222	30.50707	1.16E-07	3.098391
Within Groups	33.41667	20	1.670833			
Total	186.3333	23				

This output came from Excel, but it's important, and not too difficult, to obtain one by hand using a hand-held scientific calculator:

The columns are always Source, SS, df, MS, and F. The letter F is for Ronald Fisher, who developed this technique early in the 20th century; it represents the variance ratio. Some people use the terms "F ratio," and "mean square ratio" synonymously. The rows are always Between, Within, and Total.

Excel includes the entries for p-value and F_{crit}. Remember from earlier in this unit that the p value is the probability of getting a less likely outcome; as before, a p value less than alpha corresponds to rejecting the null hypothesis, assuming a 5% significance level. Our p value, expressed as a decimal, is 0.000000116, so we have no second thoughts about rejecting. Such a small p value means that our outcome—that is, our variance ratio—is extremely rare.

Note also that since Excel reports F_{crit} within the ANOVA table, it was an unnecessary extra step to compute =finv(). But it didn't hurt us.

Continuing toward the left, we have the individual mean squares, the degrees of freedom, and the sums of squares. You've probably realized by now that the most difficult part of any ANOVA problem is the calculation of the sums of squares: the degrees of freedom can be read off directly from the table, and the rest of the entries involve simple division. On tests and quizzes, I usually give my students the values for the sums of squares and ask them to complete the rest of the ANOVA table and answer interpretive questions, although on homework assignments I do ask them to compute the sums of squares.

Also note that the sums of squares total to a quantity called SS_{total} and similarly for the degrees of freedom. The mean squares do not, of course, because they are averages. There is no such thing as "MS_{total}."

The effect size, symbolized by η^2, is the strength of the effect that was the subject of the ANOVA. It is calculated as the ratio $SS_{between} / SS_{total}$. In our above experiment, η^2 = 0.821. The effect size can never be greater than 1, since $SS_{between}$ is one of the two terms that add to the total. One can have a strong effect and high or low p value, and similarly a weak effect and high or low p value.

Example 1

Suppose a pop quiz is given to three categories of high school students: those who were advised to get up early, eat a full breakfast, and work out (category A); those who typically pull all-nighters, skip breakfast, and do not work out (category B); and a control group who received no advice. The completion times for nine students in all three categories were recorded. Please state an appropriate null and alternative hypothesis; by hand, compute the ANOVA table; report the effect size and state the conclusion. Use the facts that $SS_{between}$ = 753.9 and SS_{within} = 277.3.

The null hypothesis is always that the population means are equal: H_0: $\mu_A = \mu_B = \mu_C$.

The alternative hypothesis is the claim that at least one μ differs from the others.

Initially, our ANOVA table would be set up as follows:

Source	SS	df	MS	F
Between	753.9			
Within	277.3			
Total				

First, the effect size: $SS_{total} = SS_{between} + SS_{within} = 1031.2$, so $\eta^2 = 753.9/1031.2 = 0.73$.

Next, the shape of the table determines $df_{between}$ and df_{within}. Since there are three samples, $df_{between}$ = 3-1 = 2, and since each of the three samples contains nine observations, df_{within} = (9-1) + (9-1) + (9-1) = 24.

Source	SS	df	MS	F
Between	753.9	**2**		
Within	277.3	**24**		
Total	**1031.2**	**26**		

Each mean square is the ratio of its SS divided by its df, and the F ratio is $MS_{between}/MS_{within}$:

Source	SS	df	MS	F
Between	753.9	2	376.9	**32.6**
Within	277.3	24	11.6	
Total	1031.2	26		

Since we're doing this problem by hand, we'd better use Excel to report the critical number:

finv(.05, 2, 24)	4.625016

Since the test statistic exceeds the critical number, we reject the hypothesis of equal population means, which is as we hoped.

ANOVA with the graphing calculator

On the TI-83, we can perform ANOVA this way: First, enter your lists of data by choosing Stat – Edit. The calculator can handle up to 20 columns. Then choose Stat – Tests, then press the keys ALPHA followed by COS. Input your lists by choosing 2nd, then 1 (for L1), 2nd then 2 (for L2)) and conclude by pressing Enter.

The Tukey Post-Hoc Test

In both of our examples, the evidence led us to reject the null hypothesis, which means that at least one μ differs from the others. But something seems incomplete: Exactly which μ, or μs, differ(s) from which others? ANOVA does not answer this question, but we would hardly feel satisfied without shedding some light.

The post-ANOVA analysis of why H_0 was rejected—assuming it was rejected—has been the subject of a great amount of statistical study. Many statisticians have proposed techniques for determining which sample(s) are sufficiently extreme as to warrant rejection of the null. The simplest one was developed by John Tukey. Tukey's test does not have the best statistical properties, but it's adequate and very easy for a beginner to understand.

Tukey's test does require same-sized samples, apropos of the caveats at the beginning of this unit. ANOVA doesn't have this requirement, but technically the Tukey test is not part of ANOVA; it's the technique one chooses after finishing the ANOVA, specifically when one has rejected the ANOVA null hypothesis of equal population means.

It goes like this: Compute the differences between all pairs of sample means, and also compute another threshold, which Tukey called the Honest Significant Difference (HSD). Any pair whose difference exceeds the HSD represents a pair that contributes to the rejection of the null hypothesis.

Although it operates like a critical number, HSD is computed—not looked up—as follows:

$HSD = q\sqrt{\frac{MSwithin}{n}}$ where MS_{within} comes from the ANOVA table and n is the common sample size. The term q, called the Studentized Residual, must be looked up in a chart.

Example

Continuing the above example, we will need the means of the individual samples:

SUMMARY				
Groups	*Count*	*Sum*	*Average*	*Variance*
A	9	89	9.888889	3.361111
B	9	202	22.44444	18.02778
C	9	121	13.44444	13.27778

The first step in our Tukey post-hoc test should be to compute all the pairwise sample mean differences. To simplify our work, we observe that we only need differences in one direction, since $|\overline{x_A} - \overline{x_B}| = |\overline{x_B} - \overline{x_A}|$. In other words, we can cut our work in half by taking absolute values of the differences and restricting ourselves to the upper right triangle of the table:

Table of mean differences

	A	B	C
A	0	12.6	3.6
B		0	9
C			0

The diagonal consists of zeroes, as it must since the difference of a number from itself is always zero. The bottom triangle contains three empty cells since they contain no information that isn't already in the upper triangle: the difference between the means of A and B is the same as the difference between B and A, except for direction.

Thus, sample means A and B differ by 12.6; A and C differ by 3.6; and B and C differ by 9 (up to absolute value). We already know that the null hypothesis of equal population means was rejected; whichever of these pairs exceeds the HSD will tell us why.

It can be shown that q, the studentized range statistic, equals 3.53. We also have MS_{within} = 11.6 and n = 9; then in that case we would have

$$\text{HSD} = 3.53\sqrt{\frac{11.6}{9}} = 4.008$$

Any pair whose difference exceeds 4.008 would contribute to the rejection of the null. In our example, those pairs are A with B and B with C. In other words, B is off the scale relative to A and C. We actually could have guessed this, since $\overline{x_B} = 22.4$ is so much greater than the means for A and C (9.9 and 13.4). What we didn't know before, but is now clear having performed the Tukey post-hoc test, was whether A and C were sufficiently far away from each other to also contribute to the rejection of the null: they are not.

Example 2
A nearby stretch of road has a posted speed limit of 55 mph, but we suspect that actual motorist speeds might vary by time of day. A random sample of speeds is recorded as follows:

Morning	Afternoon	Evening	Overnight
60	60	56	61
54	61	54	62
57	59	58	59
56	62	55	57
54	57	59	56
55	55	54	55
57	57	55	58

We will use ANOVA to test the null hypothesis that the population means are equal against the alternative that at least one (possibly more) differs:

$$H_0 : \mu_{Morning} = \mu_{Afternoon} = \mu_{Evening} = \mu_{Overnight}$$

Using the above formulas and techniques, we obtain this ANOVA table:

ANOVA						
Source of Variation	*SS*	*df*	*MS*	*F*	*P-value*	*F crit*
Between Groups	44.67857	3	14.89286	2.823928	0.060173	3.008787
Within Groups	126.5714	24	5.27381			
Total	171.25	27				

Since the p value is greater than the default alpha level of 0.05, we fail to reject the null hypothesis. Based on our sample, we cannot reject the claim that the population means are equal. We would obtain the same conclusion by noticing that the test statistic 2.8239 is less than the critical value 3.0087.

The between variance estimate is 14.892; the within estimate is 5.27. These estimates are quite different, but not different enough to count against H_0. Their ratio is 2.8239; the ratio would have to exceed 3.008 in order to count against H_0.

Note that the various degrees of freedom are $df_{Between} = 3$, since there are 4 treatment levels; $df_{Total} = 27$, since there are 28 observations in total; and $df_{Within} = 24$, which accords with the experimental design, since total number of observations – number of treatment levels = 28 – 4 = 24.

Since we did not reject the null hypothesis, we may possibly be making a Type II error: Maybe H_0 really is false, but we didn't detect this because of biases in our sample Finally, we skip the Tukey procedure: Tukey is only performed when the null is rejected, in order to shed light on which combinations of samples are making us reject, but we did not reject, so there is no light to shed.

Two-Way ANOVA

Let's return to the coffee shop example:

A	B	C	D
4.5	5	9	2
3	5	11	5
3	4	9	6
5	4	11	7.5
6	5.5	12	6
4.5	4	11	5

Our One-Way ANOVA experiment asked whether these service times, grouped by stores A, B, C, and D, came from populations with the same means or whether at least one store had a different population mean service time. Suppose we now ask whether service time, in addition to being grouped by store, can be grouped by some other factor: new barista versus veteran, in-store versus drive-through, whether the patron orders food versus whether the order only consists of drinks, and so on.

We'd ordinarily suppose that we would need an entirely new set of data if we wanted to test two null/alternative pair of hypotheses. It turns out, though, that with careful arrangement of the data, we can indeed test two null/alternative pairs on the same data. There are some restrictions, but the gain in efficiency may be worth the extra effort.

The technique is called Two-Way ANOVA, the "two" referring to the cross-classification of the data. We will use the same table as before:

	A	B	C	D
Veteran barista	4.5	5	9	2
	3	5	11	5
	3	4	9	6
Newbie barista	5	4	11	7.5
	6	5.5	12	6
	4.5	4	11	5

I've chosen to cross-classify by barista type: veteran or newbie. There are 12 mixed orders (three each for A, B, C, and D) and also 12 drink-only orders. The intersections between column and row treatment levels are called **interaction blocks**, and it is critical that these interaction blocks all have the same size. If they don't, your software will not be able to perform the test.

The observations within each interaction block are called replicated; hence, technically the full title for this procedure is **Two-Way ANOVA with Replication**. We imagine that three measurements were made on mixed orders at store A; three on drink-only orders at store A; etc.

Also critical is that you, the experimenter, have carefully matched the observations with the corresponding interaction block. This wasn't such an issue with One-Way ANOVA, since we only had columns to worry about. But with interaction blocks, we cannot just randomly assign observations to columns. As designed above, the three drink-only orders must be recorded below the three mixed orders.

The reason to perform a Two-Way ANOVA with Replication (we'll call it Two-Way ANOVA for short) is that we can test two null hypotheses on one data set:

For the columns, we have as before

$H_0 : \mu_A = \mu_B = \mu_C = \mu_D$ H_1: at least one μ differs.

Similarly, for the rows—although Excel calls them "the sample"—we have

$H_0 : \mu_{mixed} = \mu_{drink}$ and H_1 that they differ. (When there are only two factors, "at least one differs" means the same as "they differ.")

However, the calculations for the column ANOVA from above will change, probably considerably, when introducing a second factor. It's not as simple as affixing the ANOVA for the factor to the previous results. Everything must be recalculated.

Even more important, in fact the first thing to check, is whether there is an interaction. Interactions are evil; if your data has an interaction, the entire Two-Way ANOVA procedure must be stopped. It makes no sense to perform a Two-Way ANOVA in the presence of an interaction.

Why? Basically, an interaction represents a third main effect. We designed this experiment under the assumption that there was a main effect due to store (the first H_0) and a main effect due to order type (the second H_0). An interaction may well obscure the main effects, making our inferences dangerously questionable.

Mathematically, an interaction is based on the means of the interaction blocks:

Interaction Block Means				
	A	B	C	D
Veteran	3.5	4.666667	9.666667	4.333333
Newbie	5.167	4.5	11.33333	6.166667

The mean of the three replicated in the A-mixed interaction block is 3.5, of the B-drink only interaction block 4.5, and so on. All the entries in this table are means of corresponding interaction blocks.

The case we desire, of no interaction, occurs whenever all the values in a row treatment level retain their rank order across the columns. If all the row values—Veteran A, Veteran B, Veteran C, and Veteran D—were lower (or all were higher) than the corresponding values in the Newbie row (drink A, drink B, etc.), then we would say there is no interaction and would proceed with the analysis of the main effects.

As it happens, unfortunately, store B spoils the pattern. Veteran is lower than Newbie for A, C, and D, but Newbie is lower than for B. That's exactly what we didn't want: for the interaction block means to flip rank order across the columns. This is the definition of an interaction, and the experiment must be halted. We will not be able to analyze the main effects.

Fig 2.8 shows an interactions plot: a line graph of the interaction block means. The four column treatment levels are indicated on the x-axis, and the y-axis shows the scale on which the interaction block averages are measured (from 3.5 up to 11.33). A polygonal line is drawn for Veterans, and another one for Newbies. Since the means flip rank order across the columns, the polygonal lines cross each other. Thus an interactions plot gives visual confirmation of what we already knew: there is an interaction, and we must not proceed to an analysis of the main effects.

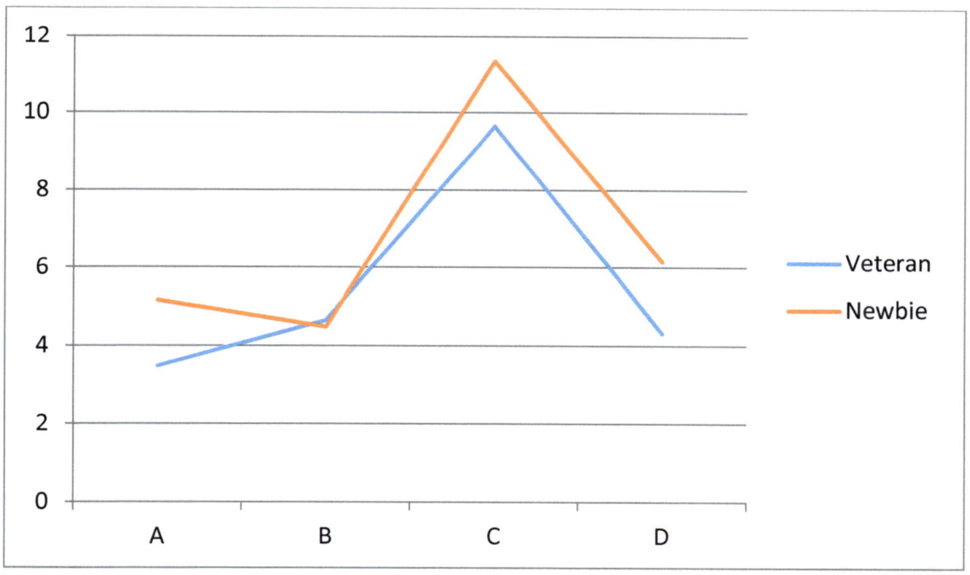

FIG. 2.8 INTERACTIONS PLOT OF VETERAN VS. NEWBIE BARISTAS

Consider this short example of a Two-Way ANOVA that has no interaction:

	A	B	C
	19	18	18
	21	18	16
new	24	21	16
	24	24	15
	11	11	9
	14	14	8
used	14.5	14	9
	15	15	11

We're supposing that the observations represent car prices, in thousands, and that A, B, and C are car models. The cross-classification is whether the car was bought new or used.

As an exercise, take a few minutes to produce the table of interaction block means. It should be obvious that the interaction blocks involving new cars always have higher means than the corresponding interaction blocks involving used. Sketch the interactions plot, and convince yourself that the polygonal line a will not cross, since means of new cars will always be greater than means of used cars. Hence, there is no interaction and the analysis may proceed.

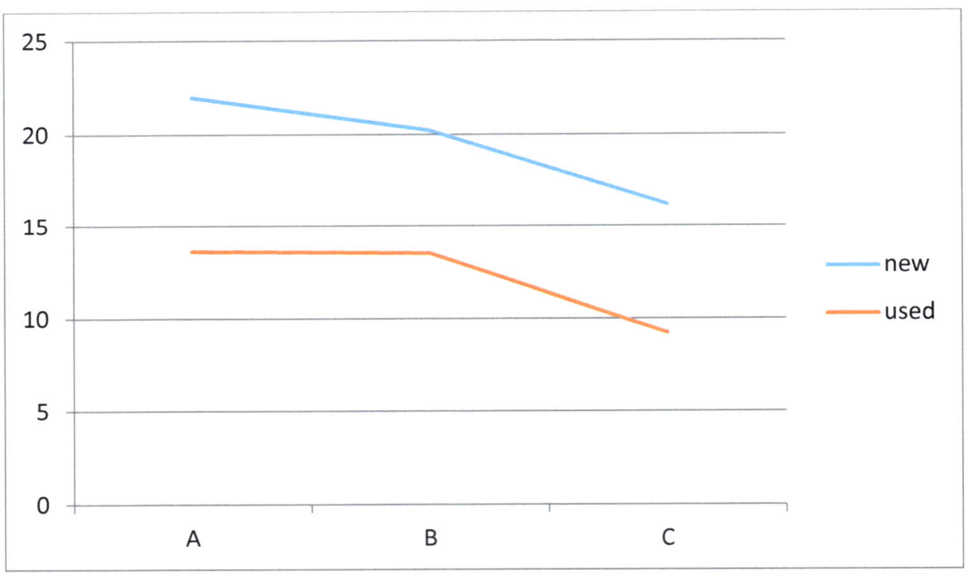

FIG. 2.9 INTERACTIONS PLOT OF NEW VS. USED CARS

Two-Way ANOVA can be performed with many applications; here is the relevant part of Excel's output:

ANOVA

Source of Variation	*SS*	*df*	*MS*	*F*	*P-value*	*F crit*
Sample	326.3438	1	326.3438	82.80793	3.74E-08	4.413873
Columns	116.0625	2	58.03125	14.72511	0.000163	3.554557
Interaction	3.0625	2	1.53125	0.388546	0.683593	3.554557
Within	70.9375	18	3.940972			
Total	516.4063	23				

Remember, there are two main null hypotheses, each with its own p value and F ratios (technically there are three nulls because Excel tests the interaction, but we may pass over this). For the sample—that's what Excel calls the row factor—the p value, in decimal notation, is 0.0000000374. This is quite a bit below the threshold of 0.05, so we may reject the claim that new cars are worth the same as used cars. And the same with car model: p = 0.000163, which is less than 0.05, so we reject the claim that the population mean car models are equal. This is the ideal situation: no interactions, and we reject the nulls of equal population means on both of the main effects.

Since there are two levels of the row variable, $df_{sample} = 2 - 1 = 1$; since there are 3 levels of the column variable, $df_{columns} = 3 - 1 - 2$. We compute $df_{interaction}$ as the product of the previous two: $df_{interaction} = df_{sample} \times df_{columns} = 2 \times 1 = 2$. df_{Total} is one less than the total number of observations in the entire experiment (24 – 1 = 23), and df_{Within} is the total in the experiment minus the product of the number of row and column treatment levels: 18 = 24 – 2 x 3.
As always, MS = SS / df, but observe we have four mean squares instead of two, as with One Way ANOVA. But, as with One Way ANOVA, there is no MS_{Total}.

With Two Way ANOVA an F ratio is computed with the main effect in the numerator, and MS_{Within} as the denominator:

$$F_{Sample} = MS_{Sample} / MS_{Within}$$
$$F_{Columns} = MS_{Columns} / MS_{Within}$$
$$F_{Interaction} = MS_{Interaction} / MS_{Within}$$

although we would already know whether or not there is an interaction because we would have produced the interactions plot before arriving at this point in our analysis. Indeed, if there had been an interaction—if the polygonal lines had crossed—we would not be analyzing the main effects at all.

A word about the sums of squares: Computing these can be complicated enough in One Way ANOVA; for Two Way, I believe they seriously obscure the main ideas, so I do not recommend demonstrating these for beginning students. It's enough to understand that SS_{Sample}, SS_{Column}, $SS_{Interaction}$, and SS_{Within} add to give SS_{Total}, and that SS = MS × df for all four lines in the output screen.

Let's have one more Two Way example:

Suppose we wish to test the effectiveness of a new drug, as measured by the duration in hours of symptoms. Let's design a Two Way ANOVA so that we can cross-classify both by gender and drug type. A placebo, by the way, is an inert drug with no medicinal benefits; it allows us to establish a baseline for comparison.

	Female	Male
Placebo	90	91
	87	91
	102	98
Drug A	90	91
	80	88
	75	85
Drug B	89	87
	72	60
	69	72

We have two column treatment levels, three row treatment levels, six interaction blocks and three replicates per interaction block.

The null hypothesis for the columns is $H_0 : \mu_{Female} = \mu_{Male}$ with the alternative that they differ. The null for the rows is $H_0 : \mu_{Placebo} = \mu_{DrugA} = \mu_{DrugB}$ with the alternative that at least one μ differs.
We'd better test for an interaction:

The table of interaction block means:

	Female	Male
Placebo	93	93.33333
Drug A	81.66667	88
Drug B	76.66667	73

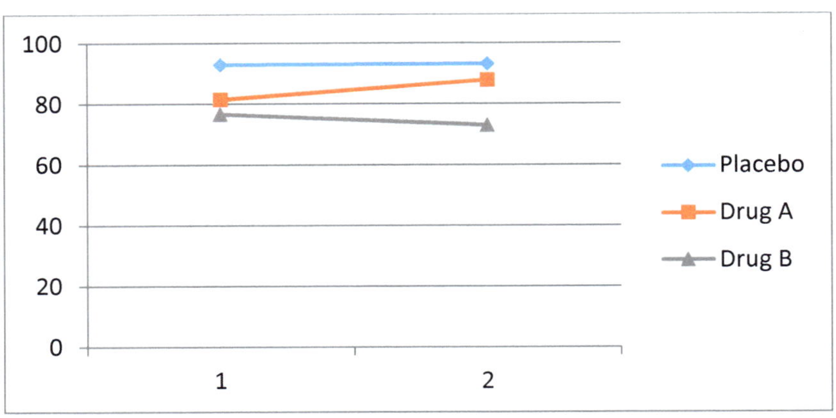

FIG. 2.10 INTERACTIONS PLOT OF DRUG EFFECTIVENESS BY GENDER (1 FEMALE, 2 MALE)

Since the polygonal lines don't cross, there is no interaction. The duration of symptoms is longest for the placebo, regardless of gender; Drug A has the second-longest symptom duration, regardless of gender; and Drug 3 has the shortest duration for both symptoms.

Note that, although the lines do not cross, they are not parallel. On some interpretations, this is sufficient for an interaction, so check with your software or research supervisor to determine how stringently to interpret this.

For our purposes, since the lines don't cross we will proceed to the main effects:

ANOVA						
Source of Variation	*SS*	*df*	*MS*	*F*	*P-value*	*F crit*
Sample	1011.111	2	505.5556	6.801196	0.010602	3.885294
Columns	4.5	1	4.5	0.060538	0.809806	4.747225
Interaction	76	2	38	0.511211	0.612262	3.885294
Within	892	12	74.33333			
Total	1983.611	17				

Check that SS / df = MS for Sample (Rows), Columns, Interaction and Within. The sums of squares total to 1983.611, and the degrees of freedom total to 18. There are 3 row treatment levels, so df_{sample} = 3-1 = 2; there are 2 column treatment levels, so df_{column} = 2-1 = 1; the interaction df = 2×1 = 2. There are 18 observations in the table, so df_{Total} = 18-1 = 17. And since there are 2 column treatment levels and 3 row treatment levels, df_{within} = 18-2×3 = 12.

MS equals SS / df for all rows of the table except for Total, and the F ratios equal MS for the main effects divided by MS_{Within}. In the end, though, we really want to know about the two null hypotheses: Since the p value for sample (rows) is less than the alpha level of 0.05, we reject the claim that the drugs have equal effectiveness. But since the p value for the columns exceeds alpha, we cannot conclude men and women suffer different duration of symptoms.

References

Erceg-Horn, D. and Vikki M. Mirosevich (2008). "Modern Robust Statistical Methods: An Easy Way to Maximize the Accuracy and Power of Your Research," *American Psychologist* 63: 591-601.

Frankfort-Nachmias and Anna Leon-Guerrero (2014). *Social Statistics for a Diverse Society*, 7th ed. Los Angeles: Sage.

Gonzalez, R. (2009). *Data Analysis for Experimental Design*. New York: Guilford.

Gravetter, Frederick J. and Larry B. Wellnau. (2004). *Statistics for the Behavioral Sciences*. Belmont: Wadsworth/Thompson.

Henkel, R. (1976). *Tests of Significance*. Beverly Hills and London: Sage.

Iverson, G. and Helmut Norpoth (1976). *Analysis of Variance*. Beverly Hills: Sage Publications.

Kennedy, P. (1994). *A Guide to Econometrics*. Cambridge, MA: MIT Press.

Hubbard, R. and M. J. Bayarri (2003). "Confusion Over Measures of Evidence (p's) Versus Errors (α's) in Classical Statistical Testing," *The American Statistician*, 57, 171-182.

Lohr, Sharon (1999). *Sampling: Design and Analysis*. Pacific Grove: Duxbury.

Rosnow, R. and Robert Rosenthal (1989). "Definition and Interpretation of Interaction Effects," *Psychological Bulletin*, 105, 143-146.

Saville, D. J. (1990). "Multiple Comparison Procedures: The Practical Solution," *The American Statistician*, 44, 174-180.

Smithson, M. (2003). *Confidence Intervals*. Beverly Hills: Sage Publications.

Snedecor, George W. and William G. Cochran (1989). *Statistical Methods*, 8th ed. Ames: Iowa State University Press.

Stuart, A. (1984) *The Ideas of Sampling*. New York: Oxford University Press.

Sudman, S. (1976) *Applied Sampling*. San Diego: Academic Press.

Thompson, Bruce (2006). *Foundations of Behavioral Statistics: An Insight-Based Approach*. New York: Guilford.

Triola, Mario (2014). *Elementary Statistics*, 12th ed. Boston: Pearson.

Chapter 2 Glossary

Alternative hypothesis	Denoted H_1, that which we are claiming about μ when the null hypothesis is rejected.
ANOVA	Analysis of Variance. One Way ANOVA (or Single Factor ANOVA) is a test of the null hypothesis that multiple population means are equal, with the alternative that at least one population mean differs. Two Way ANOVA is similar except that the data are cross classified, allowing for the simultaneous tests of two such nulls (and their alternatives).
β	β is the probability of a Type II error, computed as β = 1 – power.
Central Limit Theorem	Sample means are normally distributed with mean μ and standard deviation $\frac{\sigma}{\sqrt{n}}$
Confidence interval	A range on the x-axis that we believe contains the population mean. A 95% confidence interval means that we believe 95% of such intervals would contain μ under conditions of resampling.
Critical number	The boundary of the rejection region.
Degrees of freedom	The number of free choices for the unknown parameter, subject to certain fixed constraints. For the t distribution, df = n-1. For One Way ANOVA, $df_{Between}$ = # treatment levels – 1 and df_{Within} = total # observations – # treatment levels.
Effect size	In One Way ANOVA, denoted η^2, this is computed as $\frac{SS_{Between}}{SS_{Toral}}$.
HSD	Honest Significant Difference. It is used in the Tukey test.
Interaction block	In Two Way ANOVA, an interaction block consists of the replicated in each combination or row and column treatment levels.
Mean Square	Same as variance. MS is always a sum of squares divided by an integer interpreted as

	degrees of freesom. $MS_{Between} = \frac{SS_{Between}}{df_{Between}} and MS_{Within} = \frac{SS_{Within}}{df_{Within}}$
Null hypothesis	Denoted H_0, this is a claim about an unknown population parameter, often μ.
P value	The probability of getting an outcome that is less likely than the outcome you actually got, assuming the null hypothesis is true. Equivalently, the p value is the area beneath the distribution curve beyond the test statistic that corresponds to the arrow indicated in H_1, or twice that area if H_1 is stated with ≠.
Power	The probability of correctly rejecting the null hypothesis.
Rejection region	The area beneath the distribution curve beyond the critical value(s).
Significance	Denoted α, this quantity equals 1 – confidence.
Test statistic	The quantity that is computed from the data (e.g., z score, t score, F ratio and [as we will see in chapter 4] calculated chi-squared).
Tukey post-hoc test	When the null is rejected in One Way ANOVA, the Tukey procedure is one among many tests to shed light on which sample, or combination of samples, is responsible for the rejection.
Two Way ANOVA	When data are cross-classified into rows and columns, such that the interaction blocks have equal numbers of replicates, it is the simultaneous test of equal population means on the row and column factors.
Type I error	The mistake of rejecting a true null hypothesis.
Type II error	The mistake of failing to reject a null hypothesis that is false.

Chapter 2 Sample Test #1

1. (12 pts)

a. Please use a 95% confidence interval to estimate μ if
$\bar{x}$ = 31.0, s=3.5, and n=49.
b. Please recompute the 95% confidence interval using the same sample mean and sample standard deviation as in a., but this time let n = 2,401.
c. What is the benefit of using bigger sample sizes to compute confidence intervals?
d. Please interpret the number 95%.

2. (12 pts)

a. Please test the claim H_0: μ=75 against the two-sided alternative using the default significance level. Use sample data $\bar{x}$ =75.5, s=4, n=64.
b. What type of error could you be making?
c. Please compute the probability of correctly rejecting H_0 if actually μ=75.1.
d. Please rework parts a. and b. using a sample size n=16.

3. (30 pts) On your answer sheet you will find a partial One-Way ANOVA table. Assume 18 total observations. Please fill in the blanks, and then answer these questions:

a. Please state H_0, H1, and your conclusion.
b. Is the Tukey post-hoc procedure called for? Justify your answer. If it is called for, please perform it on the back of the answer sheet.
c. Please draw and label an appropriate graph.
d. What is the probability of getting a variance ratio that is bigger than the ratio you actually computed?
e. What number is the biggest value the calculated variance ratio could have while still supporting H_0?
f. What number is the area beneath the graph beyond 4.493998?
g. What number is the area beneath the graph beyond 1.574883?
h. Assuming equal-sized samples, how many samples were used in this experiment?

4. (12 pts) This graph is associated with a Two-Factor with Replication ANOVA.

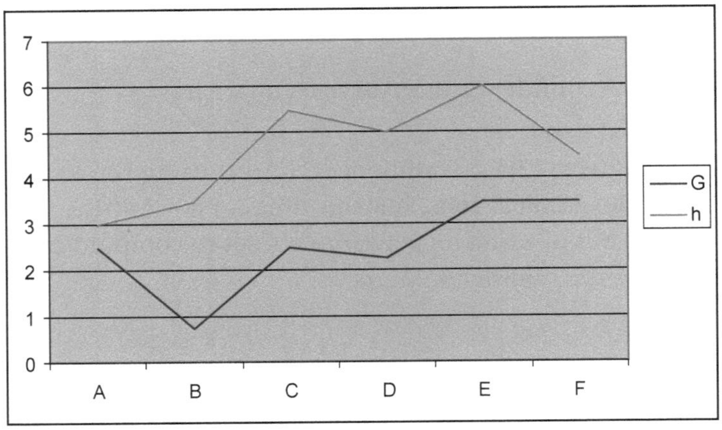

a. Please explain what information this graph tells you. Define any statistical terms you use. Be explicit but as brief as possible.
b. Based on the above graph, please compute df columns, df sample, and df interaction.
c. We discussed an important condition that, if satisfied, would make us stop such an ANOVA, and if not satisfied, would allow us to continue the ANOVA. What is this condition; is it satisfied for the data associated with this graph; and please suggest whether we would continue the ANOVA.

5. (15 pts) On your answer sheet you will find a partial Two-Way without Replication ANOVA table. Assuming four levels of the column variable and two levels of the row variable, please compute the missing entries on the answer sheet and state the null and alternative hypotheses.

6. (24 pts) Please write the entire word TRUE or the entire word FALSE on the answer sheet.

a. According to the Central Limit Theorem, each of the sample means equals μ.
b. According to the Central Limit Theorem, the distribution of sample means is normal only if the population has a normal distribution.
c. If H_0 is false, then one cannot commit a Type II error.
d. If H_0 is true, then one is bound to commit a Type I error.
e. If the sample size is less than 30, then one must use the T distribution for hypothesis tests.
f. The power of a hypothesis test equals $1 - \beta$, where β equals the probability of a Type II error.
g. The alternative hypothesis in One-Way ANOVA is the claim that no two population means are equal.
h. The alternative hypothesis in One-Way ANOVA is the claim that at least one sample mean differs from the other sample means.

Chapter 2 Sample Test #2

1. (30 pts) A One-Way ANOVA was performed on three samples, each with five observations. The variance estimate based on each observation was reported as 0.90, while the variance estimate based on the sample means and grand mean was reported to be 64.26667.

a. Use this information to produce the ANOVA table.
b. Excel reports the following:

finv 3.885

Use this information to report your conclusion to the experiment in APA style, please.

c. Please report η^2.
d. Please state the null and alternative hypotheses.

- **i.** Is the area under the curve beyond the test statistic greater than or less than the significance level, or has enough information been provided to determine this?
- **ii.** What is the area beyond 3.885, or do we have enough information to determine this?
- **iii.** What number is the boundary of the rejection region, or do we have enough information?
- **iv.** What is the biggest value the test statistic could have before rejecting the null hypothesis, or do we have enough information?

f. What error type is possible?
g. If the Tukey test is called for, please perform it on your answer sheet, being sure to explain which numbers are most relevant and why. In case you need it, q = 3.77. Some summary information is provided below. If Tukey is not called for, please explain.

SUMMARY

Groups	*Count*	*Sum*	*Average*	*Variance*
A	5	41	8.2	0.7
B	5	11	2.2	0.7
C	5	43	8.6	1.3

2. (10 points) Each of the following graphs corresponds to data that are intended for a hypothetical One-Way ANOVA experiment. In each case, please suggest whether ANOVA is, in fact, appropriate, and if not, explain.

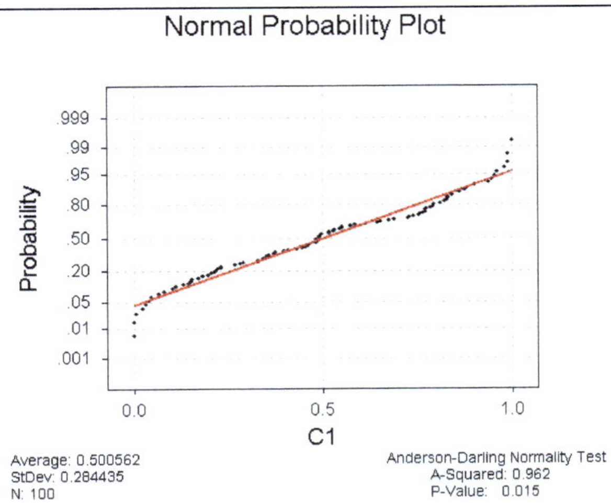

a. This graph is based on one of the samples.

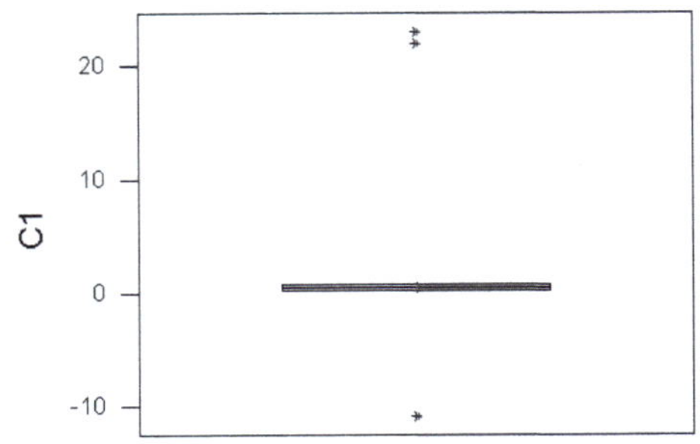

b. This graph is based on one of the samples.

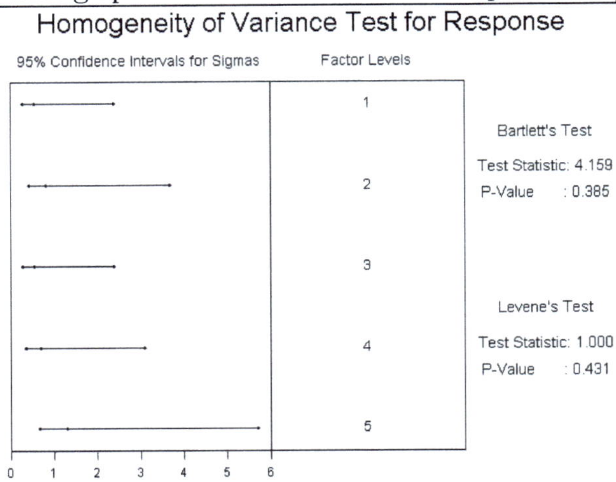

c.

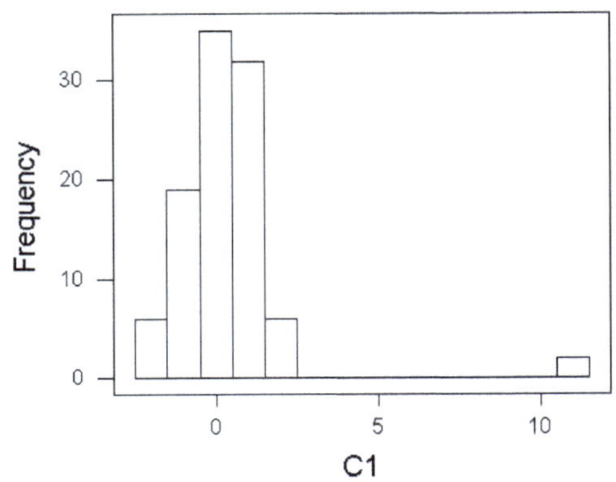

d. This graph is based on one of the samples.

3. (20 pts) Data are arranged for a Two-Factor ANOVA with Replication, given below:

	A	B	C
D	4	2	6
	6	4	8
E	1	2	4
	3	2	2

a. Please state all relevant hypotheses.
b. What must be considered before the experimenter considers an analysis of the main effects? Be explicit and show any relevant graph (sketched neatly by hand) on your answer sheet. Be sure to explain what information your graph conveys.
c. If an analysis of the main effects is called for, use the output below to report your conclusions. If not called for, please explain.

ANOVA

Source of Variation	*SS*	*df*	*MS*	*F*	*P-value*	*F crit*
Sample	21.33333	1	21.33333	12.8	0.011675	5.987378
Columns	12.66667	2	6.333333	3.8	0.085869	5.143253
Interaction	4.666667	2	2.333333	1.4	0.316961	5.143253
Within	10	6	1.666667			
Total	48.66667	11				

4. (2 points each) Please write the entire word TRUE or FALSE on your answer sheet.

a. For any given sample size, the t distribution has bigger critical numbers than the normal.
b. Any null hypothesis can be rejected, given sufficiently big sample size.
c. As sample size increases, t critical numbers approach zero.
d. As sample size increases, normal critical numbers approach zero.
e. For any hypothesis test, if $p < \alpha$, the test statistic will exceed the critical value.
f. For any hypothesis test, if H_0 is true, a Type II error is possible.
g. For One-Way ANOVA, if η^2 is close to 1, then p will be less than α.
h. For Two-Way ANOVA with replication, $df_{interaction} = df_{sample} \times df_{column}$.
i. $SS_{between}$ is based on the sample means and the grand mean.
j. The t distribution is appropriate only if the data are highly non-normal.

5. (3 pts each)

a. Define the term *significant*, as in "The results of the experiment were *significant*."
b. Define the term *power*, as in "the *power* of the test against a particular alternative."
c. Define the term *robust*, as in "The t distribution is *robust* against departures from normality."
d. Define the term *p value* in terms of "probability of an event …"
e. Define the terms *Type I error* and *Type II error*.

6. (3 pts each) Does exposure to lead affect IQ scores? An important study involved children who lived within 7 km (about 4 miles) of a large ore smelter in El Paso, Texas.

a. Based on measured blood lead level, the children were partitioned into groups. Please identify these groups.
b. What statistical technique is appropriate for this experiment?
c. Did you reject H_0 or fail to reject it?

7. (10 pts) Please use the diagram to report the power of the test and the probability of a Type II error.

Chapter 2 Sample Test #3

1. Which one of the following is NOT true regarding sampling?

  a. A sample is a subset of the population
  b. Sampling is performed when the population parameter is already known
  c. Bias is to be avoided
  d. Samples should be well-randomized

2. Please select the true statement about the Central Limit Theorem.

  a. The CLT assumes that a fixed number of samples has been collected, each with infinite sample size
  b. The CLT says that the standard deviation of the curve of means equals σ
  c. The CLT says that sample means has a t distribution
  d. The CLT says that the mean of the sample means equals μ

3. Which statement is false regarding confidence intervals for μ?

  a. For fixed n, 90% intervals are longer than 99% intervals
  b. For fixed n, there will be a trade-off between confidence and precision
  c. 95% of all such intervals will contain μ (assuming the default confidence level)
  d. If n increases, there is no trade-off between confidence and precision

4. Choose the false true statement about the t distribution.

  a. The t is more conservative than the normal.
  b. For the t distribution, df = n – 1 (n is sample size)
  c. As df increases, t critical numbers approach normal critical numbers
  d. The t curve is right-skewed with an intercept at the origin

5. Why is the t distribution used much more widely than the normal, even with big sample sizes?

  a. Because t distribution tables are easier to read
  b. Because t critical numbers are bigger than normal critical numbers for same-sized n
  c. Because t scores are easier to compute than z scores
  d. Because the t curve has nicer properties than the normal curve

6. The p value is (choose the false statement) ...

a. The probability of getting a more extreme outcome
b. The probability of getting a less likely outcome
c. The area beyond the critical value
d. Less than 0.05 when the null is rejected

7. Choose the false statement.

a. If $p < \alpha$, a Type I error is possible
b. If the test statistic is bigger (in absolute value) than the critical number, the null will be rejected
c. A Type I error is always made when the null is rejected
d. p(Type II error) = 1 – power

8. Choose the false statement about One-Way ANOVA

a. We desire $SS_{between}$ to be as big as possible, relative to SS_{within}
b. A post-hoc procedure such as Tukey's will be called for when the null is rejected
c. SS/df = MS for both between and within
d. When η^2 is big, F_{calc} will be small

9. Regarding One-Way ANOVA, if $F_{calc} < F_{critical}$, then we conclude that (choose the false statement) ...

a. No two population means are equal
b. The Tukey procedure will not be called for
c. $p > 0.05$
d. A Type II error is possible

10. Regarding One-Way ANOVA, if $SS_{between}$ is much bigger than SS_{within}, we conclude (choose the true statement)...

a. F_{calc} will be greater than $F_{critical}$
b. p will be less than 0.05
c. The Tukey procedure will not be called for
d. The effect size will be very high

11. Suppose a Two-Factor ANOVA has been performed, with this interactions plot. Choose the false statement.

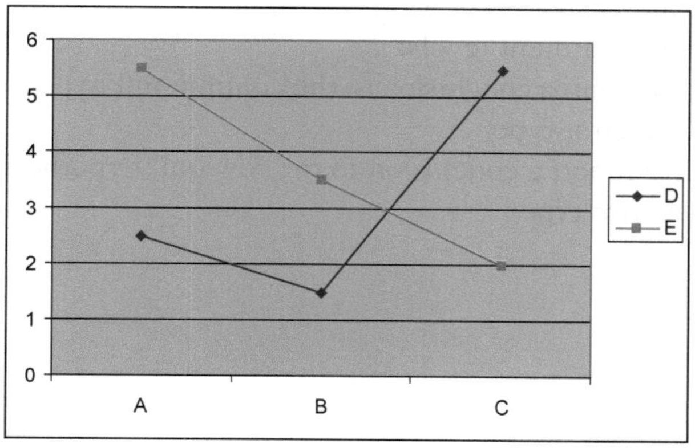

a. Reject the null hypothesis of equal column μs
b. The interaction block means have flipped rank order across the columns
c. The analysis should be abandoned
d. The interaction effect is obscuring the main effects

12. (20 pts) Suppose a One-Way ANOVA has resulted in a significant result at $\alpha = 0.05$, $F(2, 15) = 3.68$.

a. How many samples were in this experiment?
b. How many total observations were in this experiment? Assume same-sized samples.
c. Will p be bigger than or less than 0.05?
d. Which is possible, a Type I error or a Type II error?
e. What decimal is the boundary of the rejection region?
f. If Tukey is called for, please perform it using this summary data and explain. $q = 3.67$, $MS_{within} = 0.966$.

SUMMARY				
Groups	*Count*	*Sum*	*Average*	*Variance*
A	6	19	3.166667	1.366667
B	6	35	5.833333	0.566667
C	6	55	9.166667	0.966667

13. (20 pts) On the answer sheet, please report the probability of a Type II error and the power if $\beta = 0.40$.

14. (20 pts) On your answer sheet, you will find an incomplete Two-Factor ANOVA table. Please fill in the missing entries. Assume there is no interaction.

- **a.** How many column treatment levels are there?
- **b.** How many row treatment levels?
- **c.** Is it possible to report a conclusion to the column null hypothesis? If so, please state it and the possible error type.
- **d.** Is it possible to report a conclusion to the row null hypothesis? If so, please state it and the possible error type.

Chapter 2 Sample Test #4

1. (20 pts) Multiple samples were collected of computer prices, with the One-Way ANOVA table appearing below. Please write your answers on the answer sheet:

ANOVA

Source of Variation	*SS*	*df*	*MS*	*F*	*P-value*	*F crit*
Between Groups	75083.33	4	18770.83	4.961454	0.004393	2.75871
Within Groups	94583.33	25	3783.333			
Total	169666.7	29				

a. Please indicate the null and alternative hypotheses.
b. Please indicate your conclusion, and draw and label an appropriate curve, including shading the p value region.
c. Please fill in the blank: For this experiment, if the variance ratio is as big as _____ but no bigger, the data would support the null hypothesis.
d. What number is the probability of getting a variance ratio that would be less likely than the variance ratio actually obtained?
e. Assuming same-sized samples, how many samples were in this experiment? How many observations per sample? How many total observations were in this experiment?
f. What number is the boundary of the rejection region?
g. What number is the area beyond the critical number?

2. (20) Suppose an ANOVA is performed on four samples, with five observations per sample.

Assume $SS_{between} = 600$, $SS_{within} = 1600$.

a. On your answer sheet, please report the One-Way ANOVA table. Remember to report $F_{Critical}$.
b. True or False: The p value for this experiment is less than 0.05.
c. Please report the effect size η^2.
d. Please report your conclusion to this experiment using the APA style, as we discussed in class.

3. (20) An ANOVA is performed on four samples, with summary statistics as follows:

SUMMARY

Groups	*Count*	*Sum*	*Average*	*Variance*
A	3	105	35	16
B	3	132.5	44.16667	20.58333
C	3	100.5	33.5	6.75
D	3	97	32.33333	6.083333

For this experiment, the p value is 0.0123.

a. Is the Tukey procedure called for? Explain.
b. If the Tukey procedure is called for, please perform it on your answer sheet. Be sure to BRIEFLY explain what the numbers mean. If Tukey is called for, use q = 4.53.

4. (20) Consider the following data for a Two-Way ANOVA:

	A	B	C
	3	1	2
D	4	1	3
	7	5	8
	9	11	2
E	13	18	4
	12	16	1

a. Please NEATLY sketch an interactions plot on your answer sheet.
b. What does the interactions plot tell you? HINT: No partial credit will be given for merely writing "There is (is not) an interaction." As part of your answer, you will have to show that you know how to define the term "interaction" and how this idea is expressed in the graph.
c. Based on your answer to part b., what would be your recommendation concerning the Two-Way ANOVA table?

5. (10)

a. Please state the three statistical assumptions that data must satisfy in order for an ANOVA to be valid.

b. In connection with your answer to part a., please consider the following graph. Are the data appropriate for an ANOVA? Explain, but be brief.

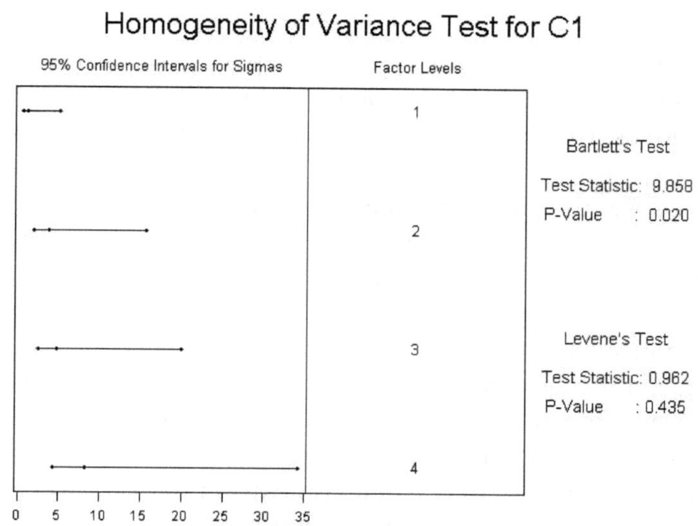

Chapter 2 Sample Test #5

1. (5 points each) For the following problems, indicate whether the appropriate statistical test would be (a) a hypothesis test using critical numbers from the normal distribution; (b) a hypothesis test using critical numbers from the t distribution; (c) One-Way ANOVA;(d) Two-Way ANOVA; (e) none of the above:

i. You have noticed a variation in cell phone prices. You suspect that one factor that causes this variation could be the model of the cell phone, so you collect a sample of 10 Trac phones, 10 Verizons, 10 Virgin Wireless, and 10 iPhones.

ii. It is claimed that the population mean amount of time college undergraduates study outside class is 24.5 hours per week. To test this claim, you survey a sample of n=13 undergraduates. This sample has no outliers.

iii. You have noticed a variation in points scored by college football teams. You suspect that one factor responsible for this variation might be conference (SEC, Pac-12, Big Ten, etc.) so you sort the data according to these categories. Assume there are 40 observations per conference. On the other hand, you also suspect that whether the team is playing home or away could also explain the variation, so you cross-classify the data according to this second treatment level.

iv. Suppose someone claims the population mean amount of time spent in line at the college cafeteria is eight minutes. To test this claim, you obtain a sample of n=22 and calculate the sample statistics. You do notice, however, that the data set is badly skewed, with more than one outlier.

2. (30 pts) The following questions refer to the output below from a Two-Way ANOVA:

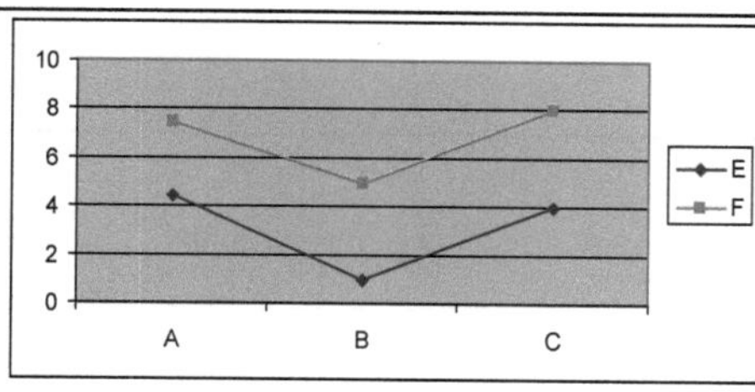

ANOVA

Source of Variation	*SS*	*df*	*MS*	*F*	*P-value*	*F crit*
	40.3333		40.3333		0.00204	
Sample	3		3		3	
Columns		2	12	8		5.143253
			0.33333	0.22222	0.80704	
Interaction		2	3	2	4	5.143253
Within	9	6	1.5			
Total						

a. What information is provided in the graph? Be brief.
b. Should the main effects be analyzed? Justify your answer.
c. How many column treatment levels are in this experiment?
d. How many row treatment levels?
e. How many observations, in total, were in this experiment?
f. How many replicates were in each interaction block?
g. If the main effects should be analyzed, please report your conclusion on the column null hypothesis.
h. If the main effects should be analyzed, please report your conclusion on the row null hypothesis.
i. If the main effects should be analyzed, will the p value for the column treatments level be less than 0.05? Justify your answer.
j. If the main effects should be analyzed, will F calc for the row treatment level be less than F critical for the row treatment level? Justify your answer.
k. Please report SS interaction. Show your work.
l. Please report SS total. Show your work.

3. (40 pts) Answer the following questions based on this ANOVA table, in which there are four samples each with six observations:

ANOVA

Source of Variation	*SS*	*df*	*MS*	*F*	*P-value*	*F crit*
Between Groups			29.26389	10.60926		3.098391
Within Groups						
Total						

a. Please report your conclusion using APA style reporting.
b. Will the area beyond 10.60926 be greater than 0.05, less than 0.05, or is it possible to determine this? Justify your answer.
c. Please report η^2, if it can be determined.
d. What is the area beyond 3.09391? Can this be determined? Round to three decimals.
e. What is the variance estimate in which each individual observation is used? Round to three decimals.
f. Consider this table of mean differences. Is the Tukey post-hoc test called for? Justify your answer. If it is called for, please perform it and make a clear connection with the above ANOVA, but be brief. Use q = 3.96.

	A	B	C	D
A	0.000	0.667	3.667	4.500
B		0.000	3.000	3.833
C			0.000	0.833
D				0.000

g. In the context of this ANOVA, please briefly but clearly explain the information that is provided in the following graph:

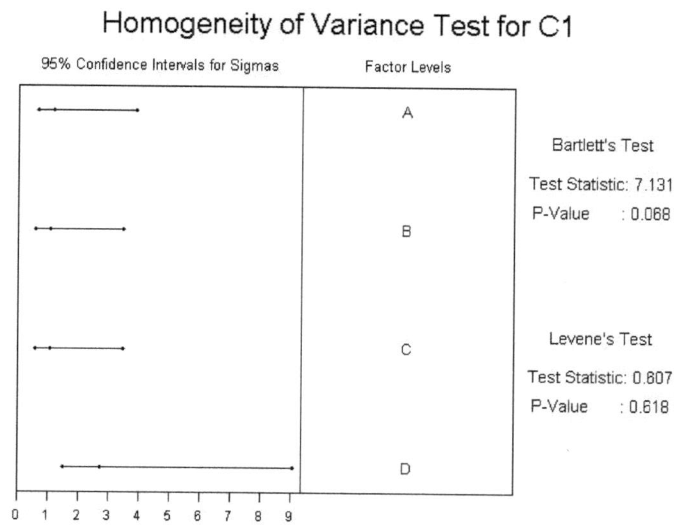

h. Please explain the Rule of Two, but be brief. What may be concluded from the Rule of Two in this experiment?

4. (15)

a. Please define the phrase "power of a hypothesis test."
b. Please complete the table on your answer sheet using the terms Type I error, Type II error, or Correct Conclusion.
c. What does it mean to say that the t distribution is more conservative than the normal?
d. Please report two ways to increase the power of a hypothesis test.
e. Which of the following are involved with the conclusion of a hypothesis test (list all that apply): μ, α, H_1, test statistic.

Chapter 3: Linear Regression

Introduction

Our next topic, single-variable linear regression, introduces us to the study of bivariate data. ANOVA used multiple samples, but they were independent and possibly of different sizes; in regression, by contrast, there will always be two samples each of the same size. The observations will not be independent, but associated pair-wise: each x value will be explicitly matched with a y value. Furthermore, the set of matched pairs will be assumed highly correlated; we will quantify this relationship with the correlation coefficient r, which is always between -1 and 1, inclusive.

To perform a regression is to calculate the line of best fit through a set of points in the Cartesian coordinate plane. The set of points is called a scatterplot, and the best-fit line is called the regression line. Although you may previously have studied many forms of the equation of a line, statistics uses the slope-intercept form y = mx + b. The slope coefficient m in particular has a useful statistical interpretation. We will be interested in the formulas for m and b, as well as r.

Since m and r are computed from sample data, they are estimates of corresponding population parameters that are unknown. The discussion follows exactly as in the development of confidence intervals and hypothesis tests for μ based on $\bar{x}$. We will use the slope coefficient m to compute a 95% confidence interval for the unknown population slope, and we will use the correlation coefficient r to perform a hypothesis test (5% significance of course) for the unknown population correlation.

Regressions are most useful for making predictions on the variable of interest (the response, or y value). The quality of the prediction can be quantified by the coefficient of determination, R^2, which is always between 0 and 1. It turns out, although we will not prove this, that the coefficient of determination is the square of the correlation coefficient.

When performing a regression, one reports a line of the form y = mx + b; to make predictions, one substitutes the desired value for the regressor x and reports the predicted value, y^. The difference between the measured y and the predicted y^ is called a ***residual***, and the sum of squared residuals, $\Sigma(y - y\wedge)^2$, will play a crucial role in our analysis.

Finally, we will address the question of when R^2 becomes "good." It can range from 0 to 1, but what is the tipping point for it to become statistically significant? Unfortunately, there is no way of answering this directly because R^2 has no distribution, unlike the t and F statistics. However, R^2 can be transformed into an F statistic via the Regression ANOVA procedure. We then will be able to answer, somewhat indirectly, that R^2 is good when F_{calc} exceeds $F_{critical}$.

The Regression Experimental Design

Let us consider bivariate data, by which I mean data sets in which pairs of measurements are made on each object of interest. Unlike the experimental design in the case of One-Way ANOVA, the pairs of measurements are not independent samples; in fact, we wish the measurements to be highly correlated, either directly (they increase together) or inversely (one increases as the other decreases). The concept of correlation is common enough in regular parlance that we may take it for granted for now; we shall quantify it mathematically soon enough.

Like ANOVA, linear regression attempts to explain the variation in a set of data; unlike ANOVA, regression achieves this by postulating an independent variable (or x variable, or regressor) whose value can be used to predict the value of an output variable (or y variable, or dependent variable). It will be very useful, since bivariate data sets have two dimensions to identify them with the Cartesian coordinate axis: an ordered pair of measurements will be interpreted as an (x, y) pair. The graph of all such (x, y) pairs is called a scatterplot, an important graphical technique for analyzing bivariate data.

Synonyms for regression variables	
X variable	**Y variable**
Independent variable	Dependent variable
Regressor	Response
Explanatory variable	Variable of interest

Examples

Example 1

Most teachers emphasize to their students that, within certain limits, more time spent studying between classes results in higher grades: a direct relationship. Imagine we have collected a random sample of, say, average hours spent studying between classes in a certain course and score on a test. Because test score is presumably the main variable of interest, we will make it the response, and the regressor will be time spent studying. A hypothetical data set and accompanying scatterplot might look like this Excel spreadsheet:

Hours studying (x)	Score (y)
1.0	69
1.5	62
1.5	73
2.0	74
3.0	77
5.0	81
6.5	83
7.0	80
9.0	85
9.5	89

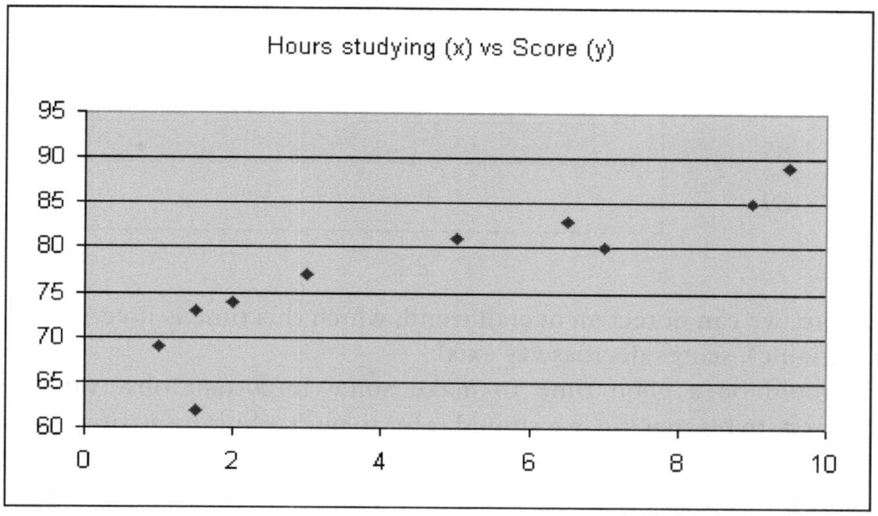

FIG. 3.1

This data set has 10 pairs of observations. The first student studied 1 hour and scored 69, the second studied 1.5 hours and scored 62, and so on. The general shape of the scatterplot trends upward, suggesting a direct relationship, although we observe a decline in the response variable between students 1 and 2 (69 decreases to 62) and students 7 and 8 (83 decreases to 80). We will apply the terms direct and inverse to a line, called the least squares line or regression line, that we will superimpose over the points in the scatterplot; this will resolve any confusion about exactly what we mean about these terms. After a few introductory comments, we will discuss the regression line in great detail.

Example 2

To continue the theme of Example 1, let us assume that if increased hours of study associate with higher test scores, then decreased leisure hours also associate with it. A bivariate dataset and scatterplot might look like this:

Leisure time (x)	Score (y)
15	69
14	62
13	73
11	74
10.5	77
6.0	81
5.0	83
5.0	80
4.5	85
2.0	89

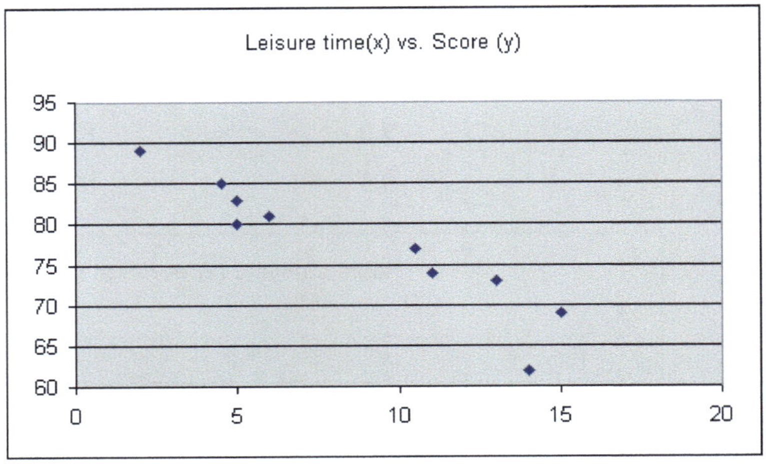

FIG. 3.2

As before, we can detect an overall trend, which this time is inverse: as leisure time increases (x variable), scores decrease (y axis).

This would be a good time to make some important observations about our assumptions. First, to be specific, we should refer to our technique as <u>single-variable</u> linear regression because we are going to restrict ourselves to one x value. It is possible to modify our model to allow two or more regressors; such models are called <u>multiple regression</u> models. If you think about it, a multiple regression model would make sense here: surely there are more independent variables than time spent studying or leisure time that are associated with (can be used to predict) test scores (examples). However, single-variable linear regression gives us plenty to discuss, so we will leave multiple regression for a future book.

Second, it might be asked why we are restricting ourselves to <u>lines</u> to make predictions on the response variable. Very many important relationships are highly non-linear; indeed, Excel and Minitab offer options for other types of regression such as exponential, logarithmic, and polynomial. Is it possible that the linear regression model oversimplifies the relationships we are considering? Yes, this is possible. Our reason for restricting ourselves to linear regression is mainly pedagogical: Lines of the form y = mx + b are very widely studied in school, reducing the overhead, one might say, required to introduce the statistical concepts. Or, to put it the other way around, we might say that the overhead required to discuss more sophisticated types of regression is far in advance of what is taught in the pre-college levels. Besides, despite its limitations, single-variable linear regression is actually pretty good at illuminating relationships. We will therefore set our concerns aside and devote ourselves, in this book, to single-variable linear regression in the knowledge that more realistic models, both in terms of number of regressors and types of regression, are available.

Third and finally, in this book we restrict ourselves to continuous regressors and responses. A variable is continuous if it can be measured, in principle, with any decimal accuracy. Examples of continuous variables include time, width, height, and temperature. I often treat currency as a continuous variable. Strictly speaking, of course, the smallest unit of (American) currency is the penny, which cannot be subdivided; but the size of the unit is usually so small compared to the range of its measurement—salaries, for example, which can be measured in the tens or hundreds of thousands (or higher)—that this small fudge probably won't bother us.

Discrete variables can only be counted by integers, not decimals: number of cars in a household, applications for a job, or books checked out from your local library, for example. Nominal variables cannot even be quantified, such as religion, gender, state of residence, satisfied/not satisfied. We will not consider discrete or nominal variables in our regression discussion, not because they are unimportant, but again, because we will have plenty to discuss regarding continuous regressors and responses. We leave the interesting discussions of non-continuous variables to future books.

We should also state at the outset that one of the chief statistical assumptions is that the response, but not necessarily the regressor, must follow a normal distribution. We are indifferent regarding the regressor's distribution—it might be normal, or not—but the response must be normal. If it isn't, then the calculations that follow are likely to be seriously misleading. Neither Excel nor Minitab automatically check this assumption, so the experimenter must make a special point to do it.

The Sample Correlation Coefficient r

As the regressor trends upward, we wish the response also to trend, although we do not care if the response trends upward or downward. The only thing we don't want is for the response to bounce around erratically, with no discernible pattern at all.

In other words, we want the regressor and response to be correlated. We can quantify the amount of linear correlation with the sample correlation coefficient, which we label r.

We will need the following summary quantities; it would be a very good idea to compute these separately, without rounding, before substituting them into the correlation coefficient formula or other formulas. Many of the formulas in this chapter are arduous to work out by hand, although it's good for beginners; separating out some of the intermediate steps reduces the chance of making mistakes that are hard to correct due to the complexity of the calculations.

Terms used in the calculation of the sample correlation coefficient

n
$\sum x$
$\sum y$
$\sum xy$
$\sum x^2$
$(\sum x)^2$
$\sum y^2$
$(\sum y)^2$
$\overline{x}$
$\overline{y}$

It takes very little time to perform these calculations in Excel, although it's probably easiest to create user-defined functions and drag the calculations down a column.

Excel actually provides built-in functions for many of these terms; in fact, it will compute the correlation coefficient using =correl(regressor column, response column). But it's good to get to know these functions by typing them in using the above quantities, and by doing it ourselves, we control round-off errors.

The correlation coefficient r is computed this way:

$$r = \frac{n\sum xy - \sum x \sum y}{\sqrt{n\sum x^2 - (\sum x)^2}\sqrt{n\sum y^2 - (\sum y)^2}}$$

This formula looks forbidding, but there are a few patterns that aid understanding:

The binomial on the numerator appears twice in the denominator, with the changes that the first denominator binomial has only x's and the second one has only y's. Both denominator binomials are multiplied, then the square root of the product is taken.

Be careful with the coefficient n, which appears in both the numerator and the denominator. It is not distributed over both terms around the minus signs; n is only multiplied by the first term, not the second.

We will see in a few pages that the correlation coefficient shares numerators with the formula for the regression slope, further easing understanding of these important formulas.

Here is what the formula looks like with the terms substituted:

$r = \frac{10(3765.5) - (46)(773)}{\sqrt{(10(306) - 2116) \times (10(60335) - 597529)}} = 0.895$. You might check this against **=correl(a2:a11, b2:b11).**

It can be proved, though we won't do it here, that the correlation coefficient is always between -1 and 1, inclusive: $-1 \leq r \leq 1$. If the data are perfectly collinear but inverse—that is, if they sit perfectly on a line with negative slope—then the correlation is -1; if they are perfectly collinear but direct, then the correlation is 1. Neither extreme happens in real life. What would very much concern us would be if $r \approx 0$, an important topic to which we will soon devote a hypothesis test.

Computing the Regression Coefficients m and b

The regression line through the scatterplot can easily be produced with Excel: Highlight the columns of data, then choose the scatterplot option in the graphs section. Right-click any one of the scatterplot dots, choose Add Trendline, then click the boxes for Show equation and Show R^2.

We can also compute the coefficients m and b by hand for the same reason as in our discussion of the correlation coefficient: besides facilitating familiarization with the concepts, we can control round-off error. Notice that Excel rounds its slope and y intercept to four decimals. For many applications, and certainly in a beginning course, four decimals may be sufficient, but we should not take for granted that it always is so.

The formula for m resembles the formula for the correlation coefficient:

$$m = \frac{n\sum xy - \sum x \sum y}{n\sum x^2 - (\sum x)^2}$$

The formula for the y intercept b assumes that m is already available:

$$b = \overline{y} - m\overline{x}$$

For our example, using the summary statistics calculated above, we obtain

$$m = \frac{10(3765.5) - (46)(773)}{10(306) - 2116} = 2.221 \text{ and } b = 77.3 - (2.221)(4.6) = 67.1$$

Hours studying (x)	Score (y)	xy	x2	y2
1	69	69	1	4761
1.5	62	93	2.25	3844
1.5	73	109.5	2.25	5329
2	74	148	4	5476
3	77	231	9	5929
5	81	405	25	6561
6.5	83	539.5	42.25	6889
7	80	560	49	6400
9	85	765	81	7225
9.5	89	845.5	90.25	7921

sum x	46	r	0.894569
sum y	773	m	2.221398
sum xy	3765.5	b	67.08157
sum x2	306		
sum y2	60335		
(sum x)2	2116		
(sumk y)2	597529		
x bar	4.6		
y bar	77.3		

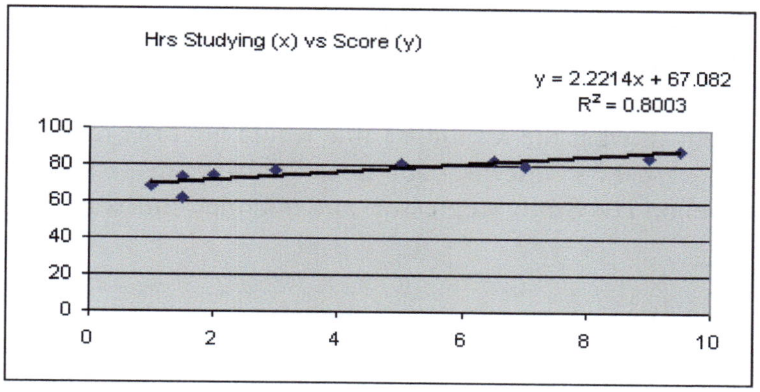

FIG 3.3 EXCEL WORKSHEET OF REGRESSION DATA

The reason for computing regression lines is to make predictions. And that's very easy: substitute the desired regressor value as x, do the arithmetic, and the answer—always labeled y^ (y hat)—is what the line thinks should be the response value associated with the regressor you plugged in.

Interpreting the Slope

The sample slope coefficient m can be interpreted as the amount of change in the response caused by a one-unit increase in the regressor. Consider this scatterplot, in which car age in months is regressed on resale value in thousands of dollars:

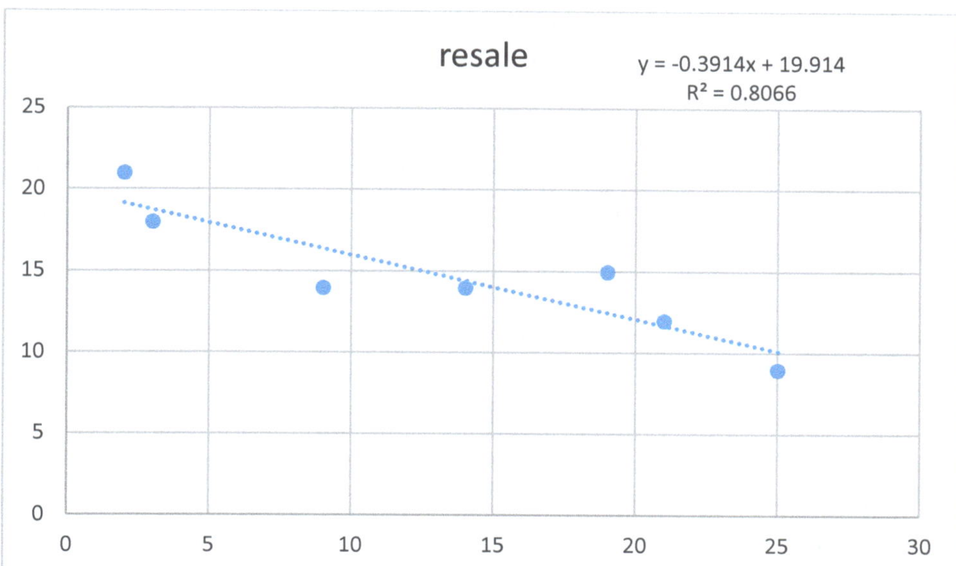

The slope coefficient m = -0.3914. This tells us that, for every extra month of the car's age, its resale value will decrease by 0.3914 thousand dollars, or $391.40.

The y-intercept b tells us what the response value would be if the regressor were zero, but be careful: This can be hard to interpret and doesn't always make sense. If we regressed Test Score (response) on Hrs Study (regressor) and obtained this scatterplot

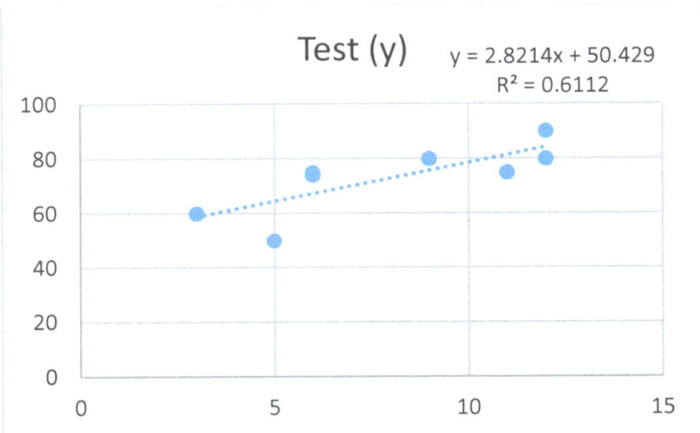

we might be tempted to conclude that someone who studies 0 hours would score 50.429. But it's just as likely that the person who didn't study learned all that was necessary during class, and scored an A. There are ways to tell the software what the response should be in case the regressor is zero, but we need not get into that here.

R^2, the Coefficient of Determination

One way to quantify the quality of our predictions is by using the coefficient of determination. It isn't difficult, but does require the background concepts Total Variation, Unexplained Variation and Explained Variation.

Total Variation is defined as the sum of the squared differences between the responses and their mean: $TV = \sum (y - \bar{y})^2$. Total Variation doesn't appear on any graph; it's used only in the analysis of regression residuals.

If the residuals are small then not much will change if we replace each y value with its associated y^. Remember that y^ is the quality we compute when we substitute the x value into the regression equation. We will denote this as EV: $EV = \sum (y\hat{}-\bar{y})^2$

Not ***much*** will change, but EV will never exactly equal TV because some y is certain to differ from its y^. In other words, there will be some nonzero residual. We will denote as UV the sum of the squared residuals: $UV = \sum (y - y\hat{})^2$

It would take a bit of work to prove this, but it turns out that EV+UV = TV. Total Variation can be decomposed into a part that depends on the regression (EV) and another that depends on the residuals (UV). Because all three quantities are nonzero, it follows that neither EV nor UV can exceed TV (if either one could, the other would be negative).

R^2 is defined as the ratio EV/TV. We often say—you should memorize this—that R^2 *is the proportion of Total Variation that is explained by the regression.*

Because all the quantities are nonnegative, we have the important property that $0 \leq R^2 \leq 1$.

Software packages, including Minitab and Excel, will automatically report R^2.

The following spreadsheet example may make these concepts more concrete:

	Age in months (x)	Resale in thousands (y)	TV terms	y hat	EV terms	UV terms
	2	21	62.55372	16.9234	14.68799	16.61867
	4	17	15.28099	16.3168	10.40637	0.466762
	5	15	3.644628	16.0135	8.541538	1.027182
	7	14	0.826446	15.4069	5.363814	1.979368
	9	15	3.644628	14.8003	2.922017	0.03988
	10	15	3.644628	14.497	1.977092	0.253009
	12	12	1.190083	13.8904	0.639186	3.573612
	17	11	4.371901	12.3739	0.514102	1.887601
	21	9	16.73554	11.1607	3.725707	4.668624
	35	8	25.91736	6.9145	38.14803	1.17831
	39	7	37.09917	5.7013	54.60632	1.686622
y bar	13.09090909					
TV	174.9090909					
EV	141.5321658					
UV	33.37963855					
R2	0.809175584					
R2 adj	0.790093143					

The value of TV, 174.909, is the sum of the squares in the column labeled TV Terms, and similarly for EV and UV. Note that to compute EV, we needed to compute the y^ values by substituting the xs into the regression equation.

The coefficient of determination does have one potential drawback: By including more regressors—more columns of x variables—we can force R^2 as close to 1 as we want. It's often good, even necessary, to perform such multiple regressions (although we won't discuss it in this book), as long as the number of regressors is determined by the content area. For example, education specialists may have good reasons to stipulate that grade point average is best predicted by hours studying per week, hours of employment per week, and household income: a 3-regressor model.

The point is that it would be deeply unethical to inflate R^2 just by including more regressors that aren't called for by the model. The adjusted R^2 addresses this problem by accounting for both the number of paired observations n and the number of regressors k: $R^2adj = 1-(1-R^2)\times\frac{n-1}{n-2}$. Note that R^2 is defined only in terms of the response values, their mean, and the predicted ys; k and n are not included. But the main points are that R^2 adj is always less than R^2, and increasing the number of regressors will always decrease R^2 adj relative to the R^2 you started with.

Graphing calculator

We can use the TI-83 to report m, b, r, and R^2. Start by entering your data by choosing Stat – Edit, then typing your data into columns L1 and L2. Choose Stat – Calc, then option 4. Confirm by pressing Enter.

Population slope and correlation

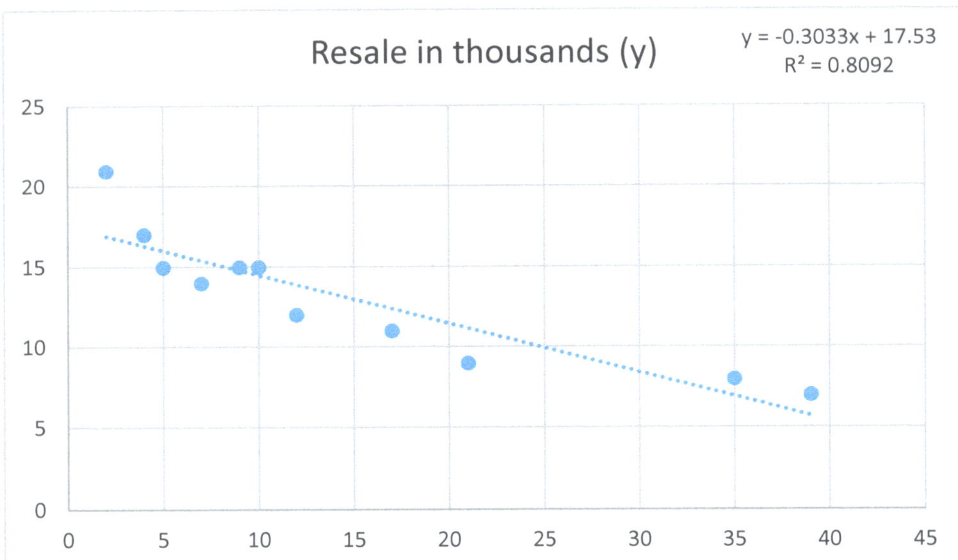

Let's discuss the relationship between the sample and population slope and correlation. We'll do the slope first. The m that we compute plays the same role in regression that $\bar{x}$ played in our work on confidence intervals and hypothesis testing; while it's useful to know the decimal value, we equally want to know what it can tell us about the unknown population parameter it estimates. The same goes for the sample correlation r and its associated unknown population parameter. Let us use Greek symbols β for the unknown population slope and ρ(rho) for the unknown population correlation:

	Slope	Correlation
Population	β	ρ
Sample	m	r

Following convention, we actually will use the data to calculate a 95% confidence interval for β, and will perform a hypothesis test for ρ.

First, to estimate the population slope: We will need the quantities m, t_{n-2}, s, and $\sum(x-\bar{x})^2$. We have already discussed how to compute m. The quantity t_{n-2} is the t critical numbers with n-2 degrees of freedom (n-2 instead of n-1 because, in regression, there are two parameters: the regression slope and intercept). The sum of squares involving the x-values and their mean might surprise us, since to this point we have been using response y-values instead, but it's not difficult to obtain them. The quantity s is the standard deviation of the data around the regression line, basically the amount of wiggle. We compute $s=\sqrt{\frac{UV}{n-2}}$.

The 95% confidence interval for β is

$$\left(m-t_{n-2}\times\frac{s}{\sqrt{\sum(x-\bar{x})^2}},\ m+t_{n-2}\times\frac{s}{\sqrt{\sum(x-\bar{x})^2}}\right)$$

Here, t_{n-2} plays the same role that the z score played earlier: it counts standard deviation units. The quantity $\frac{s}{\sqrt{\sum(x-\bar{x})^2}}$ is the standard deviation, just as, earlier, we used $\frac{s}{\sqrt{n}}$.

Before we go any further, it's important to understand how we intend to use this formula: We are mainly concerned about whether it contains zero. Just as the sample mean can vary with resampling, so can m. This varying will sketch out an interval that we believe will contain the population parameter. The crucial point is that, even if our computed value of m is not zero, the associated interval for β might well contain zero. Remember: Any point in a confidence interval is equally likely to be the value of the population parameter. If 0 is in the interval, then regardless of the decimal value for m, we cannot rule out the possibility that β may be zero. The point of regression is to show how the response changes for different regressor values. If the population slope is zero, then the response will not change with the regressor. Our work will have been wasted, futile.

It's traditional to hypothesize about the unknown population correlation. The test statistic follows approximately, though not quite, a t-distribution; I will refer to it as t*. Its formula is $t^*=\frac{r}{\sqrt{\frac{1-r^2}{n-2}}}$ and we will use the critical value t_{n-2}. The null hypothesis is the claim that the population correlation is not zero: H_0: ρ = 0 against the alternative H_1: ρ ≠ 0, for the same reason that we gave above for hoping that the confidence interval for β does not contain 0.

For example, if r = 0.65 for a regression with 14 pairs of observations, then $t^* = 2.963$ and the critical number is 2.179; for such a problem we would indeed reject the claim of zero correlation.

References

Boslaugh, S. and Paul Andrew Watters (2008). *Statistics in a Nutshell.* Sebastopol, CA: O'Reilly.

Draper, N. R. and H. Smith (1966). *Applied Regression Analysis.* New York: Wiley.

Gonzalez, R. (2009). *Data Analysis for Experimental Design.* New York: Guilford.

Kennedy, P. (1994). *A Guide to Econometrics*, 3rd ed. Cambridge, MA: MIT Press.

Lind, D., William G. Marchal and Samuel A. Wathen (2005). *Statistical Techniques in Business & Economics.* Boston: McGraw-Hill.

Mosteller, Frederick and John W. Tukey (1977). *Data Analysis and Regression: A Second Course in Statistics.* Reading: Addison-Wesley.

Ostrom, Charles W. (1990). *Time Series Analysis: Regression Techniques*, 2nd ed. Newbury Park: Sage.

Studenmund, A. H. (2001). *Using Econometrics*, 4th ed. Boston: Addison Wesley Longman.

Chapter 3 Glossary

Bivariate data	Observations are called bivariate when they are identified by ordered pairs.
Correlation coefficient	Denoted r, this statistic quantifies the degree of collinearity in the bivariate data.
R^2 (same as coefficient of determination)	The proportion of Total Variation that is explained by the regression.
R^2 adj	A modification of R^2 based on the number of paired observations and the number of regressors.
Regressor	The x-value or independent variable.
Response	The y-value, or dependent variable.
Scatterplot	A graph of the paired observations in the Cartesian coordinate plane.
β	The unknown population slope coefficient (not the same as p(Type II error) used in chapter 2).
ρ	The unknown population correlation.
Y^	The predicted response value, computed by substituting the numerical regressor value into the regression equation.
m	The sample slope coefficient, which gives the change in the response that is associated with a 1-unit change in the regressor.
b	The sample y-intercept, which equals the value of the response when the regressor equals zero.

Chapter 3 Sample Test #1

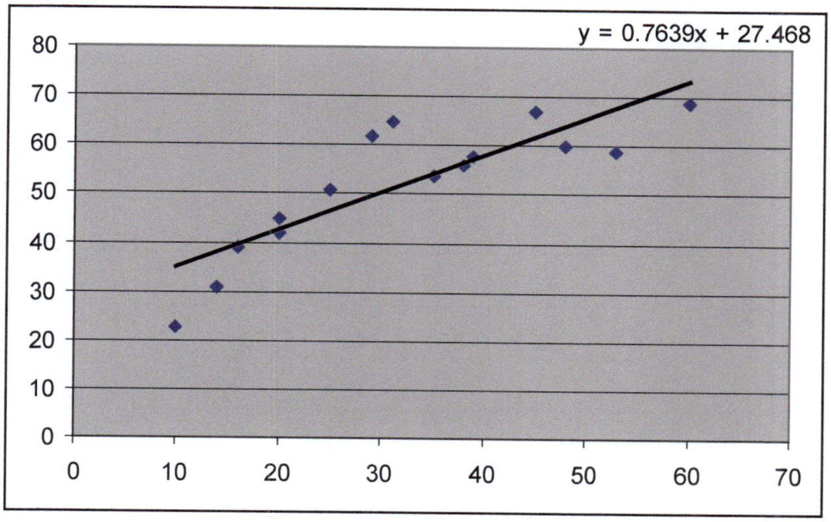

The above scatterplot refers to an experiment in which minutes spent watching television (x) are regressed on minutes spent on social media (y). Each question is worth 12 points.

1. Please predict the minutes spent on social media for someone who watches television for 30 minutes.

2. What would be the effect on minutes spent on social media, on average, of an additional minute watching television?

3. What proportion of Total Variation is explained by minutes spent watching television?

Use the facts that EV = 1840.729, TV = 2552.933.

4. Use your answer from question 3 to compute the correlation coefficient.

5. Use your answer from question 3 to compute R^2 adj. Use the fact that there are 15 pairs of observations and one regressor.

6. Please compute the Regression ANOVA table on your answer sheet. Remember to state the null hypothesis and your conclusion. The critical number for this experiment is 4.667 and the p value is 0.000062.

7. Please state the null hypothesis and conclusion for the test of the significance of the correlation. The critical number for this test is 2.16 and the test statistic is 5.80.

8. What information is conveyed by the following Residuals Plot? Be specific but do not ramble.

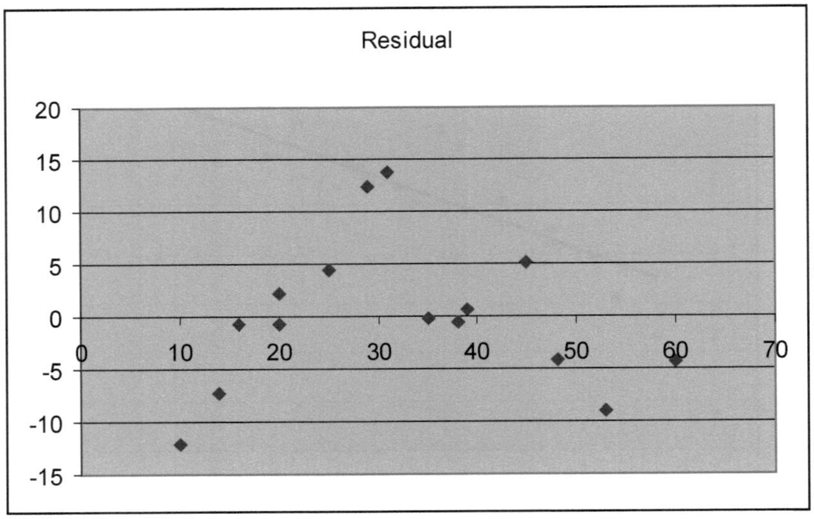

9. The 95% confidence interval for β is (0.479, 1.049). Please comment on the importance of this fact. Be specific but don't ramble.

Chapter 3 Sample Test #2

Suppose you are exploring the effect of car price on number of miles driven per year, that is, do owners of more expensive cars drive more miles per year than owners of less expensive cars? Consider the following scatterplot and summary statistics:

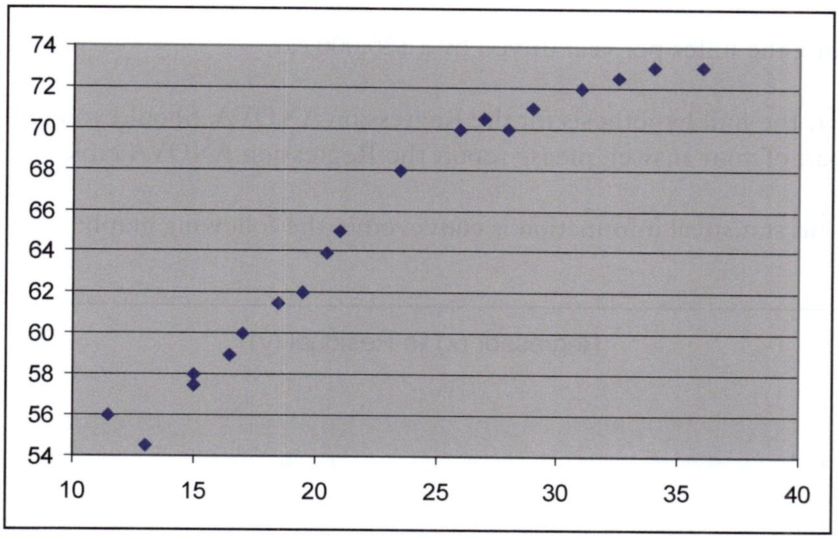

- Car price, measured in thousands of dollars, is the regressor.
- Miles per year, measured in thousands of miles, is the response.
- There were 19 pairs of observations in this experiment. In the table below, t_{17} and $f_{1,17}$ are critical numbers.

$\bar{x}$	22.87
$\bar{y}$	65.13
s_x	7.50
s_y	6.40
r	0.97
t_{17}	2.11
$F_{1,17}$	4.45
$\Sigma(x - \bar{x})^2$	1011.42
UV	38.08
TV	737.92

1. What would be the effect on miles per year of an extra thousand dollars of car price?

2. What proportion of Total Variation is explained by car price?

3. Please test for the significance of the population correlation ρ.

4. Please report a 95% confidence interval for the population slope β.

5. Please predict the miles per year driven by a $20,000 car.

6. Please report the null hypothesis for the Regression ANOVA. Should you reject it or fail to reject it? As part of your answer, please report the Regression ANOVA table.

7. What relevant statistical information is conveyed in the following graph:

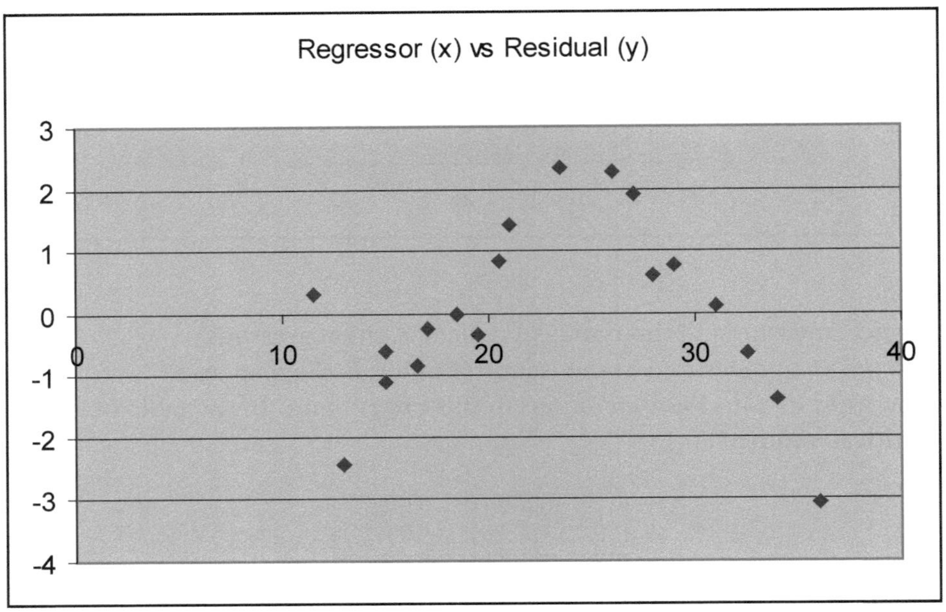

[2]

8. What relevant statistical information is conveyed in the following graph:

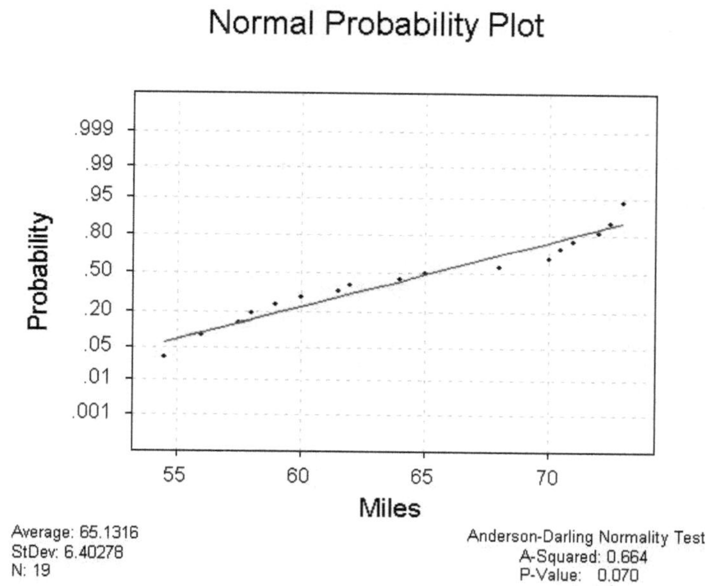

9. Which of the following quantities is/are NOT involved in the calculation of a prediction interval for y that:

n, R^2 adj, $\bar{x}$, x_0, t_{n-2}, F_{calc}, UV, Σx^2

Chapter 3 Sample Test#3

1. (6 pts) Consider the regression equation

CELL PHONE MINS = 3.5*COST + 25

where CELL PHONE MINS measures the number of minutes per month and COST is the dollar amount paid for the cell phone.

a. How many cell phone minutes are estimated to be used for someone who paid $95.00 for the cell phone?
b. What would be the effect on CELL PHONE MINS of a one-dollar increase in COST?

2. (12 pts)

a. If a regression experiment has correlation coefficient $r = 0.75$, what percent of total variation is explained by the regression?
b. If 85% of total variation is explained by the regression, and there is an inverse relationship between y and x, then what is the correlation coefficient?
c. If 30% of the total variation is ***not*** explained by the regression, then what is the correlation coefficient? Assume a direct relationship between x and y.

3. (18pts) On your answer sheet, please compute the indicated quantity or explain why it is impossible:

a. UV=100, TV=200, find R^2
b. EV=50, UV=75, find TV
c. TV=150, EV=50, find r (assume a direct relationship)
d. $r = 0.75$, TV = 250, find UV
e. $r = 0.55$, n=99, k=3, find R^2 adj
f. r=0.80, find F

4. Please use these quantities to compute the regression line and R^2 (20 pts):

n	8
Σ x	55.5000
Σ x2	445.2500
Σ y	76.5000
Σ y2	898.2500
Σ xy	439.0000
$\bar{x}$	6.9375
$\bar{y}$	9.5625

5. Please compute R^2 adj, the correlation coefficient r, and F calc (from the regression ANOVA) (20 pts):

EV	TV
8.05163	6.0025
6.248686	4.6225
2.209177	2.7225
1.319116	4.6225
1.319116	1.1025
0.04108	0.0025
1.478871	0.2025
2.414576	0.7225
4.970649	28.6225
15.3547	8.1225

6. Suppose for a particular regression that EV=75 and UV=35. Also suppose there are 11pairs of observations. Use this information to produce the Regression ANOVA table. Verify the relationship between F and R^2. (10 pts)

7. Suppose a regression of $ spent on gas (y) on miles driven to a gas station (x) produces the equation y = 25.75. Note that m = 0. What amount of $ spent on gas is predicted when x = 10 mi? x = 100 mi? How does y behave for every one-mile increase in x? (5 pts)

8. Please write the entire word TRUE or FALSE on your answer sheet (2 points each).

- **a.** The correlation coefficient r must be between 0 and 1.
- **b.** The slope coefficient m must be between -1 and 1.
- **c.** When R^2 is close to 1, F will tend to be small.
- **d.** A Residuals Plot is used to test the normality of the response.
- **e.** A Residuals Plot is used to test the normality of the regressor.
- **f.** R^2 adj is never greater than R^2.

Chapter 4: Discrete Probability

Introduction

In the first three chapters, we focused on lists of continuous observations when we discussed descriptive statistics, confidence intervals for μ, hypothesis tests for μ, One-Way ANOVA, Two-Way ANOVA, and regression. We now completely shift gears and examine events from experiments whose outcomes can be counted in terms of 0 successes, 1 success, 2 successes, and so on. We will typically use the symbol k to count the number of successes in such experiment; thus, if X is the random variable that counts the number of successes, we will be interested in computing p(X=k) where k would be among the whole numbers.

Definitions, Examples, and Properties

The concept of probability has come up before; we understood probability as the area beneath a distribution curve, in particular the normal, t, or F distribution curves. (We will introduce a fourth such distribution in this chapter: the Chi Squared.) This way of thinking about probability is useful and correct as far as it goes, but there is another way to think about it.

In this chapter, we will introduce a mathematical definition of probability that does not rely on the framework of distributions. Instead, we will define the probability of an individual event and then discuss a number of laws and properties of probability. This "discrete" approach will be fully consistent with the "distributional" approach developed earlier, although justifying the equivalence between the two approaches is beyond the scope of this book.

First, let us establish some notation: The letters E and F will stand for the events under consideration, S (called the sample space) will represent all the events that are possible, even if they did not occur, and p(E) and p(F) will denote the probabilities of E and F. Events E and F are called disjoint if they have no elements in common, that is, an empty intersection. The symbols n(E), n(F), and n(S) will stand for the number of ways E, F, and S could happen. Then we will define p(E) = n(E)/n(S).

If the sample space S is simple, then computing probabilities is a matter of routine arithmetic.

Examples

Example 1

Toss a fair, two-sided coin. Then S has two elements, H and T. If the event E is tossing a tail, we would note that there is only one way E can occur since there is only a single tail in the sample space. Thus p(E) = n(E)/n(S) = ½.

Example 2

Toss a fair, six-sided die. (The singular is "die," plural is "dice.") The sample space S consists of the six elements {1, 2, 3, 4, 5, 6}.If E is the event "getting a number greater than 1," then p(E) = n(E)/n(S) = 5/6.

Card problems can be particularly useful both in demonstrating concepts of probability and in modeling real-world probabilistic experiments, so we should describe how a deck is arranged. There are 52 cards in a deck, half red and half black. We do not use jokers. The red cards are divided into the suits of hearts and diamonds, and the black cards are spades and clubs. Thus, there are 26 red cards and 26 black cards, and each color has two suits of 13 cards each. A deck also can be cross-classified into values: in increasing order, 2 through 10, then jack, queen, king, and ace. The four jacks, four queens, and four kings are the 12 face cards. The aces have the highest values, though they are not face cards.

Example 3

Pick a card from a deck. Let E be the event "red king." Since there are two red kings, the king of diamonds and the king of hearts, we have p(E) = n(E)/n(S) = 2/52.

Notice that we did not bother to reduce the answer, 2/52, to lowest terms, nor have we converted any of our answers into decimals. Keeping answers in unreduced form has the advantage of reminding us what we took for n(E) and n(S), if at some point in the future we questioned these terms. But this isn't a very big point.

Example 4

If we pick a card from a deck and the event E is "black face card," then since there are six black face cards (spades and clubs each of jacks, queens, and kings), we would compute

p(E) = n(E)/n(S) = 6/52.

One fundamental property of probability becomes apparent from the previous examples: the probability of an event is always a decimal between 0 and 1, inclusive: $0 \leq p(E) \leq 1$. Events that have probability 1 are certain to happen, and events with probability 0 are impossible.

This fundamental property is based on the relationship between E and S. Remember that the sample space S is the set of all possible outcomes; thus, E is a subset of S. As such, E can have fewer events than S—in fact, E can be empty—and if E=S, they would of course have the same number of elements; but E can never have more events than S. It follows that the numerator of the fraction n(E)/n(S) can never exceed the denominator. If E is empty, then of course n(E) = 0 n(E)/n(S) = 0 too.

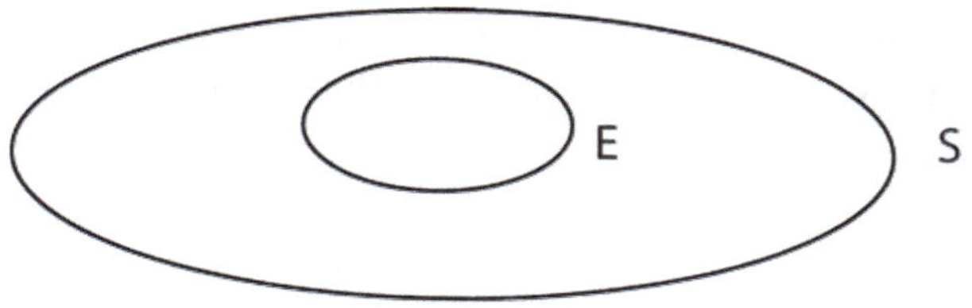

FIGURE 4.1 VENN DIAGRAM OF E AND S

Example 5

Toss a die. If E is the event "greater than or equal to 1," then p(E) = 1 because E and S have all the same elements. This is an example of an event that is certain to happen.

Example 6

If we toss a die and let E be the event "multiple of 7," we have n(E) = 0 since there are no multiples of 7 on a die. Therefore, E is impossible: p(E) = 0.

We have been computing probabilities, not odds. Probabilities and odds both are ways to quantify random events, but they have quite different properties.

Mathematically, odds(E) = $\frac{n(E)}{n(S)-n(E)}$; in other words, the odds of an event is the ratio of the ways E can happen to the ways it can fail to happen. Odds and probability have the same numerators but different denominators, with one implication that the odds of an event, unlike probability, can be greater than 1. In the media, odds are far more commonly presented than probabilities. For the remainder of this book, we will focus almost exclusively on probabilities.

If two disjoint events E and F have the properties that their union is the sample space S, then E and F are called complements. That is, E and F are called complements if they have no elements in common but, taken together, they encompass the set of all possible outcomes. The complements rule states that complementary probabilities add to 1:

If E and F are complements, then p(E) + p(F) = 1 or, equivalently, p(E) = 1 – p(F).

This result can sometimes be surprisingly useful as an indirect way to compute probabilities. Suppose we desire p(E), but this is difficult to obtain directly. If we're lucky and E has a simple complement F, then we can obtain p(E) by computing the probability of its complement and subtracting from 1. We hasten to point out that there is no guarantee that E will have a simple complement. In such cases we might seek alternative methods, possibly involving approximations. But it's always a good idea to first check the applicability of the complements rule in such cases.

Conditional Probability and Bayes' Rule

Sometimes we wish, for some reason, to restrict the size of the sample space. That is, we wish to compute p(E) subject to some condition on S. The technique is called conditional probability. It's best to present this concept via further examples.

Examples

Example 7

a. Imagine picking a card from a deck, and letting E be the event "6 of clubs." There is one 6 of clubs in the deck, so p(E) = 1/52.

b. Suppose we modify the previous example: this time, let E be the event "6, given that the card is a club." We will notate this as p(6 | club). In this problem, we are taking the sample space to be all (and only) the 13 clubs, instead of the full 52-card deck. This is why we employ the terminology conditional probability: the condition for restricting the sample space is expressed by the description "club" following the vertical slash. In this reduced sample space, there is still only one 6, so p(6 | club) = 1/13.

c. By contrast with 7b. above, note that p(club | 6) = ¼ because we are restricting to the four 6s in the deck, of which there is only a single club.

To review: p(6 of clubs) = 1/52, p(6 | club) = 1/13, p(club | 6) = ¼.

Example 8

a. Roll a die. p(odd prime number) is the number of odd primes divided by 6, since the sample space has the six outcomes 1, 2, 3, 4, 5, 6. Then p(odd prime number) = 2/6, since the only two odd primes on the die are 3 and 5. (The number 2 is also a prime, since it has no factors except itself and 1, but it is of course not an odd number. In fact, it is the only even prime number.)

b. p(odd number | prime) is the probability of getting an odd number on the die, given that we are restricting the sample space to the prime numbers 2, 3, and 5. Thus, p(odd number | prime) = 2/3.

c. p(prime | odd) restricts the sample space to the three odd numbers 1, 3, and 5. The number 1 is not a prime—it is called a unit—so there are only two primes among the three odd numbers on the die, and p(prime | odd) again equals 2/3.

We have much more to discuss regarding probability, but even at this early stage we are in a position to present one of the most remarkable and useful, but controversial, results: Bayes' Theorem. It's a conditional probability result applied to multistage experiments, that is, when the result from a first-stage experiment opens up possibilities in a second stage.

Suppose 75% of respondents in a certain poll have gotten their flu shot. Of these, 10% get the flu. Of the 25% who do not get a flu shot, suppose half get the flu. We can diagram the situation this way:

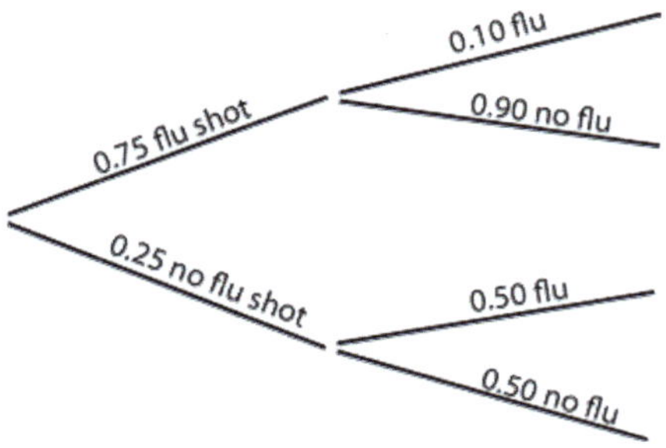

FIG. 4.2

Before we proceed to Bayes' Theorem, let's analyze the problem in a bit more generality.

Example 9

A person could get the flu if (s)he got the flu shot or did not get the shot:

p(flu) = p(shot and flu) + p(no shot and flu) = 0.75(0.10) + 0.25(0.50) = 0.20. Thus, 20% of respondents overall will get the flu.

By the Complements Rule, the probability that someone will not get the flu must be 1 – 0.20 = 0.80. This ought to correspond to the diagram, and it does:

p(no flu) = 0.75(0.90) + 0.25(0.50) = 0.80.

The above computations did not involve conditional probability, so let's deepen our analysis somewhat. The following example still does not involve Bayes, but it starts us thinking in those terms.

Example 10

Continuing with the flu shot example, the proportion of people who get the flu, given that they had the shot, is 10%: p(flu | shot) = 0.10. This is apparent since it was stipulated in the statement of the problem.

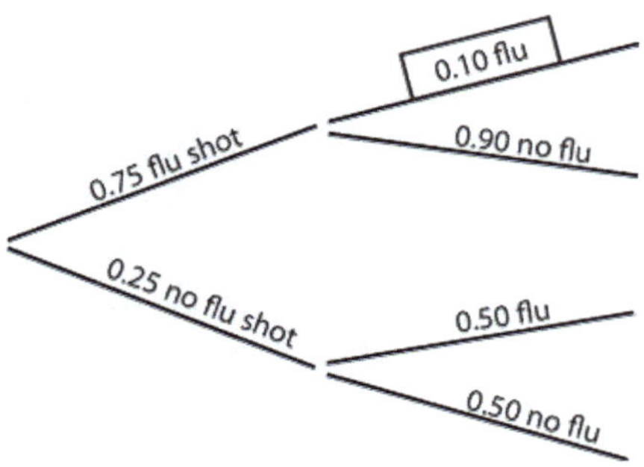

FIG. 4.3

Similarly, the proportion of people who don't get the flu, given that they haven't had the shot, is 0.50:

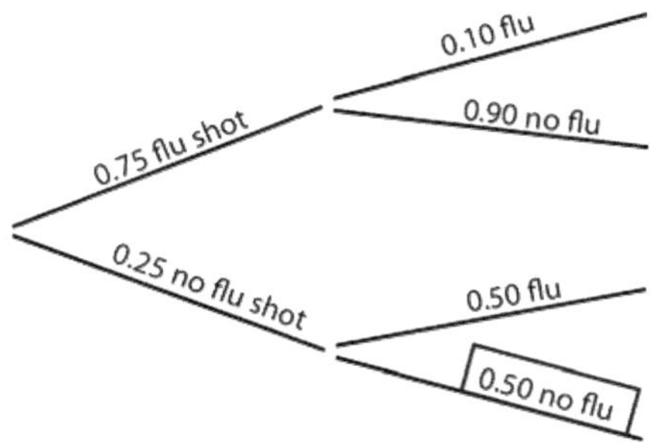

FIG. 4.4

What made Example 10 so easy, in fact so trivial, is that the condition was on the first stage of the experiment, that is, the first set of branches. In such a case, the required probability can be directly read from the second stage.

Bayes' Theorem applies in a similar case, but when the condition is applied to the second of branches. In this case, the condition is a little more complicated since it is applied over a disjoint set:

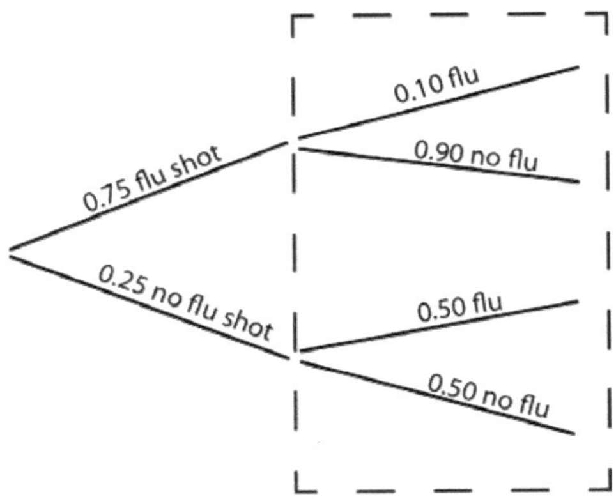

FIG. 4.5

Without providing a rigorous proof, let's consider the computations involved in such a problem. The basic difficulty is that since the condition occurs over a disjoint set, it's not immediately obvious how to represent or compute it. The key insight is to identify all paths that lead to the event described in the restriction and take the sum of such paths as our denominator. The specific event described in the statement of the problem will be our numerator.

Example 11

$$p(\text{flu shot} \mid \text{flu}) = \frac{0.75 \times 0.10}{0.75 \times 0.10 + 0.25 \times 0.50} = 0.566.$$

Thus, although 75% of all respondents got a flu shot, only about 56.7% of those who got the flu received a flu shot. When we restrict the sample space, the probability changes; or, as this is often expressed, by incorporating prior information into our model, we change the probabilities.

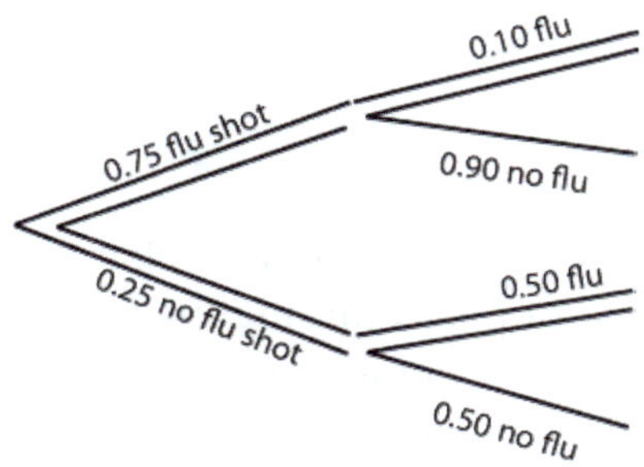

FIG. 4.6

This seems to make sense: The more information you bring to bear into a problem, or the more information you feed into it, the more accurate you expect your answer to be. Perhaps paradoxically, however, this is what makes Bayes' technique controversial, because there isn't always a guarantee that the prior information can be objectively measured. Statisticians, of all people, would not want their results to be biased by subjective assessments. We will sidestep this issue for now, focusing on the operations and interpretations.

Example 12

The probability that a client has health insurance at a certain health clinic is 0.80. Of those who have insurance, 40% have an outpatient procedure performed; of those who do not have insurance, only 5% have the procedure. What is the probability that someone has insurance, given that they did not have the procedure?

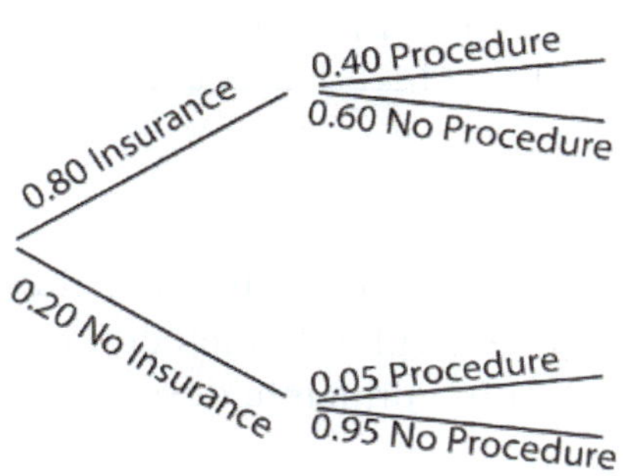

FIG. 4.7

First, we note that this problem does call for Bayes since it is a multistage conditional probability problem in which the condition is on the second stage of the experiment (i.e., having or not having the procedure).

$$p(\text{insurance} \mid \text{no procedure}) = \frac{0.80 \times 0.60}{0.80 \times 0.60 + 0.20 \times 0.95} = 0.716$$

Our initial estimate that 80% of patients have insurance declines to 71.6% if we incorporate information about having a certain procedure.

Binomial Distribution

Bayes' Theorem deals with two-stage conditional probability experiments. In this section, we will introduce a different type of problem: single-stage experiments in which we choose a subset of events from a larger set of possible outcomes. To deal with such experiments, we will develop an extremely useful model called the binomial distribution and these variations: the normal approximation, poisson, and multinomial.

The binomial distribution relies on a basic mathematical technique called the multiplication principle, and it will be very useful to state it at the beginning.

> **Multiplication Principle:** The number of ways of completing a multistage experiment is the product of the number of options at each of the stages.

Examples

Example 13

How many phone numbers are there? For simplification, assume phone numbers have 10 digits (three-digit area code, three-digit prefix, and four-digit suffix). Also assume that any digit can be used in any position. By the Multiplication Principle, the number of ways of creating a phone number is the product of the number of ways of choosing each of its 10 digits:

10	10	10	10	10	10	10	10	10	10

$10^{10} = 10{,}000{,}000{,}000$

Our simplifying assumptions are a little too simple, since there are no such numbers as 000-000-0000 or 123-456-7890, which are implicitly included in our calculation. But it's useful as an introduction to the multiplication principle despite this limitation.

Example 14

A teacher gives a 10-question True-False pop quiz. What is the probability that someone randomly guessing will get a perfect score?

By the Multiplication Principle, since there are two choices for each question, we compute 2^{10} = 1,024 ways to fill in the answer sheet.

2	2	2	2	2	2	2	2	2	2

2^{10} = 1,024. Then, assuming there is only one right answer for each question, we compute 1/1,024 as the required probability.

Example 15

A dinner at a certain restaurant requires a choice of salad out of three varieties on the menu; meat, out of four varieties; vegetable, out of four choices; dessert, out of two choices; and drink, out of five choices. By the Multiplication Principle, assuming all choices are compatible, there will be 480 dinner choices.

3	4	4	2	5

3x4x4x2x5 = 480

Because we have introduced coin and dice questions and will return to them pretty soon, let's count the elements in sample spaces that involve tossing multiple coins and dice.

Example 16

By the Multiplication Principle, there are four outcomes when we toss two dice: 2x2 = 4. These are HH, HT, TH, TT. Similarly, there are 2^3 = 8 outcomes when tossing three dice, 2^4 = 16 when tossing four, and so on.

Sample space for tossing three dice

HHH	TTT
HHT	TTH
HTH	THT
THH	HTT

Sample space for tossing four dice

HHHH	HHTT	TTHH	TTTT
HHHT	HTHT	THTH	TTTH
HHTH	HTTH	THHT	TTHT
HTHH			THTT
THHH			HTTT

Example 17

If we toss two dice, there will be 36 outcomes, shown below. There will be 216 outcomes if we toss three dice, but that's rather too many to diagram.

Sample space for tossing two dice

11	12	13	14	15	16
21	22	23	24	25	26
31	32	33	34	35	36
41	42	43	44	45	46
51	52	53	54	55	56
61	62	63	64	65	66

The symbol 13 means that we tossed 1 on the first die and 3 on the second, and similarly for the other entries.

Tree Diagrams

Why is it the Multiplication Principle instead of, say, the Addition Principle? Why do we multiply the number of choices at each stage rather than, say, add them? That's a fair question. To see why, let's list the choices for a simple problem using a graphical technique called a tree diagram. It isn't a proof, but it's usually convincing enough for the beginning student.

Suppose you've been too busy to do laundry, and your fashion choices are limited to two shirts, two pairs of pants, one pair of shoes, and three hats. By the Multiplication Principle, you should have 2x2x1x3 = 12 choices for getting dressed.

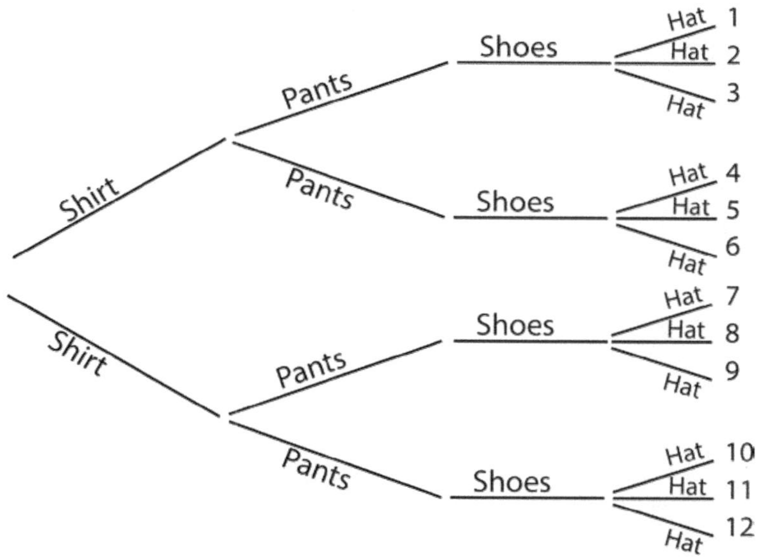

FIG. 4.8

For each of the two shirts, there are two pairs of pants; for each shirt-pants combination, there is one choice of shoes; and for each shirt-pant-shoes combination, there are three hats. You can see how the diagram branches out toward the right, hence the name "tree diagram." I've labeled the end nodes, and counting them verifies that we do indeed have 12 choices, as we predicted.

Tree diagrams are useful for justifying the multiplication rule, but they do have limitations. First, they become tedious to draw if there are very many stages. In the above problem there were shirts, pants, shoes, and hats; but if we additionally included glasses, socks, sweaters, gloves, canes, monocles, cravats, and belts, diagrams would take far longer. Second, if the number of choices at each stage is great, there will be so many branches that we may never finish drawing. Take Example 13, for example, which has 10 categories and 10 choices per category, for a total of 10,000,000,000 choices overall: a quick, back-of-the-envelope calculation shows that if we were to draw one line per second without ever stopping, it would take just about 13 years—12 years, 11 months—to complete.

That's not to say that tree diagrams are only useful for convincing people about the Multiplication Principle. Trees are used very frequently to diagram options in operations research, budgeting, and other fields. When outlining a strategy or comparing results based on competing assumptions, it's very useful to list them this way.

Let's look at one more example of a tree diagram. We'll show the sample space for tossing two dice, which as we know has 36 outcomes. We showed the sample space in a tabular form; let's see what the tree diagram looks like.

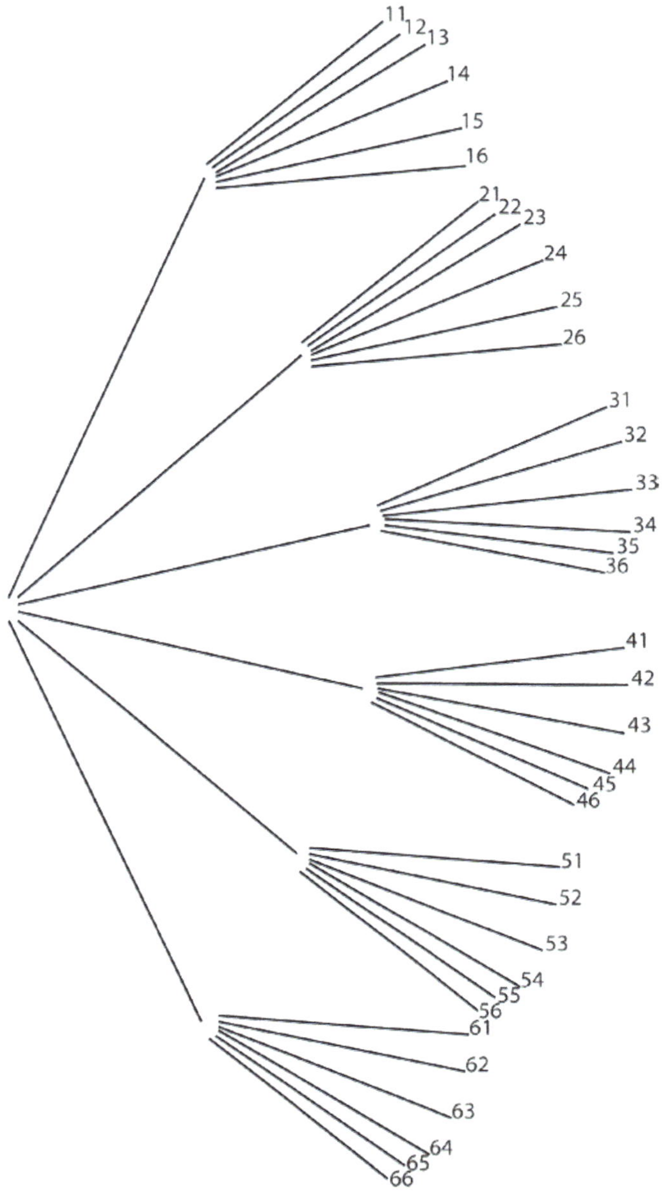

FIG. 4.9

We could give increasingly interesting and complex examples involving the Multiplication Principle, but we actually have all we need to proceed now to our main objective, which is the binomial distribution, so let us return to that.

Back to the Binomial Distribution

Let us describe a few situations in which the binomial distribution would be appropriate.

Examples

Example 18

Suppose seven people are in line at your local coffeehouse. The probability that someone pays with cash is 0.30. If X counts the number of customers who pay with cash, we might be interested in the following probabilities: $p(X=7)$, the probability that everyone uses cash; $p(X=0)$, the probability that no one does; or $p(X \geq 1)$, the probability that at least one person does.

Example 19

The probability that any individual passes her or his emissions test is 0.80. If 15 individuals bring their cars to be tested, and we let X count the number of cars that pass, we might be interested in $p(X=15)$, $p(X=0)$, or $p(X \leq 14)$.

Example 20

Imagine a certain bank investigates accusations of identity theft from five different customers. The probability of any particular occurrence of identity theft is 0.01, let us say. If X counts the number of cases of identity theft, we might consider $p(X \geq 1)$, $p(X \leq 5)$, or $p(X = 2)$.

Example 21

Some travelers pay online for their hotel rooms. Imagine that the probability that any individual pays online is 0.35. In a sample of 500 travelers, if X represents the number who pay online, we may want to compute $p(X=500)$, $p(X \geq 200)$, and $p(X < 300)$.

These four examples share a number of important characteristics that typify binomial problems, so it's worth pausing to discuss them in some detail before we start looking at formulas.

Success versus Failure

Most fundamental is that each case has only two possible outcomes, which we generically classify as success or failure: cash versus other sorts of payment, passes emissions test or fails it, identity theft or no identity theft, pay online or pay some other way. The name "binomial" comes from the fact that there are two outcomes.

"Success" isn't necessarily an outcome of which we approve, however, and "failure" isn't necessarily bad. It depends on the problem. Normal people wouldn't approve of identity theft, yet that does count in Example 15 as a success. Success is an outcome that matches what we're counting, with no judgments involved.

If a problem requires more than two outcomes, the binomial distribution cannot be used. The good news is that we can easily generalize it to a model called the multinomial distribution. But first things first.

Every binomial problem has an upper limit n on the number of possible outcomes and a random variable X to count the number of successes k where k cannot be bigger than n. The binomial theorem computes the probability of this situation. In the first part of

Example 16, for instance, n is 500 and k ranges from 200 up to 500; the question asks for the probability that the random variable X takes on these values. Keep in mind that all the rules for probability that we discussed at the beginning of this chapter remain in effect, such as $0 \leq p(X \geq 200) \leq 1$.

Perhaps it seems pedantic to point out that n is the upper limit on the number of outcomes; after all, don't all probability models have upper limits? The surprising answer is no, some do not. When we derive the Poisson distribution from the binomial later in this chapter, we will see that it assumes that the number of outcomes can approach infinity, at least in principle.

All the events must be independent. Intuitively, events are independent if the occurrence of one of them in no way affects the occurrence of the other. Mathematically, we will define events E and F to be independent if their joint probability equals the product of the individual probabilities:

If E and F are independent, p(E and F) = p(E) x p(F).

Examples

Example 22

Toss two coins and one die. Let E be the event HH on the coins, and let F be the event "greater than or equal to 5 on the die." Are E and F independent? It certainly seems so, since we can think of no way for the coins to interfere with the die. But let's apply the definition to verify this result mathematically.

p(E) = ¼, since there are four outcomes in the sample space for tossing two coins (HH, HT, TH, TT) and exactly one of them matches the description HH. We also compute p(F) = 2/6 since there are two ways (5, 6) to satisfy the description out of the six elements in the sample space. Thus, E and F will be independent if p(E and F) equals ¼ x 2/6 = 2/24.

The joint sample space consists of all 24 events in which the first two outcomes are heads or tails and the third is a digit from 1 to 6:

HH1	HT1	TH1	TT1
HH2	HT2	TH2	TT2
HH3	HT3	TH3	TT3
HH4	HT4	TH4	TT4
HH5	HT5	TH5	TT5
HH6	HT6	TH6	TT6

The two bolded events are the ones that satisfy the condition "HH on the coins and at least 5 on the coin." Therefore, we see that indeed p(E and F) = 2/24, which does equal p(E) x p(F). The two events are independent.

We made this detour because the binomial distribution requires the mathematical definition of independence. But we should note that it is a rare and restrictive property in real life. Take another look at Example 13, in which the customers at your local coffeehouse are paying with cash or some other way. Do we really believe that all customers are independent in this respect? If two customers are on a date, and one of them is paying for both, surely independence is violated: cash or plastic will be used for both orders.

Independence can be difficult to justify in real life, but the binomial distribution requires it. All we can say is that, as usual, mathematical models—especially in an introductory textbook—sometimes are only approximate descriptions of reality. They're good to start with but require considerable effort to be made more useful.

To recap: The binomial distribution applies when we have to make calculations about success or failure on a set of n objects, of which k are successes (sometimes k is stated with >, ≥, <, ≤). If the success probability is p, then by the complements rule, the failure probability will be 1-p; and assuming independence (which itself calls for the multiplication principle), the joint probability of all n objects would be p^k x $(1\text{-p})^{n\text{-}k}$. Note that the exponents add to n, as they must; if k outcomes are successes, then the remaining ones, of which there must be n-k, will be the failures.

All that remains is to address this issue: which k of the n outcomes will be the successes? This issue is not trivial, and its solution requires another detour.

Suppose we're tossing 12 dice and, letting X count the number of times we get the number 2, we desire p(X = 3). Since p(2) = 1/6, we know that the answer will involve $(1/6)^3(5/6)^9$ as shown below, where the three 2s appear first:

1/6	1/6	1/6	5/6	5/6	5/6	5/6	5/6	5/6	5/6	5/6	5/6

If the three 2s appeared last, here is what we'd get:

5/6	5/6	5/6	5/6	5/6	5/6	5/6	5/6	5/6	1/6	1/6	1/6

The product is still $(1/6)^3(5/6)^9$, of course, since none of the terms p, 1-p, n, or n-k has changed.

We now begin to see the issue: How many ways are there to choose three positions out of 12 for our 2s? The answer for each such way will be $(1/6)^3(5/6)^9$. It turns out that there will be 220 such ways, so the overall probability of getting three successes, p(X=3), will be $220\text{x}(1/6)^3\ (5/6)^9 = 0.197$, rounded; but we need to understand where the 220 comes from. Once done, we will complete our understanding of the formula for the binomial distribution and can provide answers to Examples 18, 19, 20, and 21.

Permutations and Factorials

In the previous problem there were 12 dice, of which we chose to count three 2s as successes. These successes are indistinguishable; what counts is their number, or the size of the subset, and not the order in which the 2s were chosen. Hence, we need a way to count numbers of ways to choose subsets of size k from supersets of size n. We will call it the combination of n objects taken k at a time and will notate it $\binom{n}{k}$.

How shall we compute this expression? A good approach is to divide it into two parts, first by computing the number of ways of choosing k objects from a set of n if order did matter—called the permutation of n objects taken k at a time or P(n, k)—and then turning this into a combination by accounting for the ways the subset elements could arrange among themselves. Remember: if order matters, it's a permutation; if order does not matter, it's a combination; and the binomial distribution calls for combinations.

Suppose you have a bookshelf that has space for four books and a stack of 11 books on your floor. For simplicity, we assume that all books have exactly the same shape, and all bookshelf spaces perfectly fit your books; and we will indicate the books by letters A, B, C, ... K. Also suppose you are very picky, so that the positions of the books matters to you: you will notice the difference if your books are arranged ABCD instead of BCDA or ACBD, and so on.

You can choose any one of the 11 books A, B, C ... K to fill the first position on the shelf. You make that choice by removing a book. There remain 10 choices for the second shelf position. There will be nine choices for the third position and eight for the fourth and last position:

11	**10**	**9**	**8**

By the Multiplication Principle, this problem of placing books on the shelf can be subdivided into four stages, and the numbers in the boxes represent the number of choices for each stage, hence the overall number of ways is the product of the number of choices: 11x10x9x8 = 7,920.

Because order matters, this problem qualifies as a permutation; in fact, it is P(11, 4). The p stands for permutation, 11 is the size of the set from which we are sampling, and 4 is the number of elements we are choosing. Also note that the factors decrease in steps of one.

Permutations grow very rapidly:

P(11,4)	7,920
P(11,5)	55,440
P(11,6)	332,640
P(11,7)	1,663,200
P(11,8)	6,652,800

P(11,9)	19,958,400
P(11,10)	39,916,800
P(11,11)	439,084,800

P(20, 20) is on the order of 10^{18} and P(50, 50) of 10^{64}. These are probably too big for your scientific calculator, and an expression like P(1000, 1000) is even too big for Excel. (The Excel command to compute P(n, k) is **permut(n, k).**) Fortunately, we rarely are called on to compute permutations by themselves; instead, they often occur as terms in longer expressions so that we can take advantage of canceling and other ways to simplify.

If n=k, then we have to choose all the objects from the superset, which means our expression must begin with n, then step down by one until we reach 1:

P(2,2)	2x1	2
P(3,3)	3x2x1	6
P(4,4)	4x3x2x1	24
P(5,5)	5x4x3x2x1	120
P(6,6)	6x5x4x3x2x1	720
P(7,7)	7x6x5x4x3x2x1	5,040
P(8,8)	8x7x6x5x4x3x2x1	40,320
P(9,9)	9x8x7x6x5x4x3x2x1	362,880
P(10,10)	10x9x8x7x6x5x4x3x2x1	3,628,800
P(11,11)	11x10x9x8x7x6x5x4x3x2x1	39,916,800

Such expressions, of the form P(n, n) or P(k, k), can also be written as n! or k! These are called **factorials**:

k factorial = k! = P(k, k) = kx(k-1) x (k-2)x ... x3x2x1

The Excel command is **=fact(k).**

A factorial is a special case of a permutation in which the list of factors stops at 1. Equivalently, a permutation can be thought of as a special kind of factorial that can get truncated before it hits 1. Both permutations and factorials are multiplicative lists of terms that decrease in steps of 1; when they differ, it's only because of where they stop.

P(n, 1) = n, of course, since this expression only contains the term n. Sometimes we are called on to report a value for P(n, 0), so we will define this expression to be 1. Similarly, 0! = 1. These don't quite fit the pattern we established before, because there is no list of terms that decrease in steps of one and certainly does not stop at 1. We can still make sense of these by imagining that there is only one way to choose 0 objects from a set of size n, and that one way is to choose nothing.

Combinations

Now for the second part of our analysis: We need to account for the number of ways the k objects in our subset can arrange among themselves.

Think about it: We have n objects, of which we select a subset of size k. If order did matter, as in the bookshelf example, there would be P(n, k) ways to choose the k elements. But what changes if order should not matter? The answer is that in this case the term P(n, k) overcounts by a factor of k! For a combination will not recognize the ways the k objects could permute among themselves, although a permutation does recognize this. Hence, a combination of n objects taken k at a time equals the corresponding permutation divided by the factorial of the size of the subset, or

$$\binom{n}{k} = \frac{P(n,k)}{k!}$$

Excel computes this with the formula **=combin(n, k).** An example will clarify this reasoning. Let's imagine there are five books, A, B, C, D, and E, of which we pick subsets of size 3. If order mattered, there would be P(5, 3) = 60 arrangements, and here they are:

ABC	ABD	ABE	ACD	ACE	ADE	BCD	BCE	BDE	CDE
ACB	ADB	AEB	ADC	AEC	AED	BDC	BEC	BED	CED
BAC	BAD	BAE	CAD	CAE	DAE	CBD	CBE	DBE	DCE
BCA	BDA	BEA	CDA	CEA	DEA	CDB	CEB	DEB	DEC
CAB	DAB	EAB	DAC	EAC	EAD	DBC	EBC	EBD	ECD
CBA	DBA	EBA	DCA	ECA	EDA	DCB	ECB	EDB	EDC

Notice that the 60 results have been arranged so that each of the 10 columns contains the same three books. If order does not matter, only the first row of subsets—ABC, ABD, … CDE—would contain different subsets of books. There are six entries in each column because there are 3! = 3x2x1 = 6 ways to arrange the elements in any subset of size 3. Thus, if order didn't matter, P(5, 3) would overcount the number of subsets by a factor of 3!=6.

Thus

$$\binom{n}{k} = \frac{P(n,k)}{k!}$$

Since, according to this approach, a combination is derived from a permutation, it follows that $\binom{n}{0} = 1$ for the same reason that P(n, 0) = 1. We will see soon that some problems do require us to compute terms like this.

Assumptions of the Binomial Formula

We now have discussed all the assumptions of the binomial distribution:

Success and Failure

The binomial distribution assumes that each trial shall be classified as either a success or a failure.

n Trials, k Successes

We assume that there are n trials under consideration, of which some subset k will be the successes and the remaining n-k the failures. k can be any number between 0 and n.

Success Probability p

Each trial can occur with success probability p; therefore, the probability of failure is 1-p.

Independence

The outcome of any trial does not affect the outcome of any other trial. Mathematically, this translates as p(E and F) = p(E) x p(F).

Order does not matter. The trials are considered indistinguishable, so all that matters is the number of ways to select a subset of size k from n elements, which we designate $\binom{n}{k}$ and compute as $\frac{P(n,k)}{k!}$ or $\frac{n(n-1)(n-2)...(n-k+1)}{k!}$. Note that some authors define $\binom{n}{k}$ as $\frac{n!}{k!(n-k)!}$. The two expressions are equal, but I feel that $\frac{P(n,k)}{k!}$ better expresses what a combination really is. However, when we turn to the multinomial distribution, we will call on the latter expression.

Putting everything together, we derive the result we have been waiting for:

If X is a binomial random variable, then the probability of k successes out of n trials is

$$\mathbf{P(X=k)} = \binom{n}{k}\mathbf{p^k(1-p)^{n-k}}$$

Now we'll revisit Examples 18 through 21.

In Example 18, we imagined seven customers in line at a coffeehouse, success counted as paying with cash, and p(success) = 0.30. X counts the number of successes, and we were to compute p(X=7), p(X=0), and p(X $\geq$ 1).

p(X=7) = $\binom{7}{7}(0.30)^7(0.70)^0$ = 0.0002. As a small shortcut, remember that any quantity raised to the zero power evaluates to 1. $\binom{k}{k}$ also evaluates to 1, since $\binom{k}{k} = \frac{P(k,k)}{k!} = \frac{k!}{k!}$. Thus, this problem reduces to 0.30^7.

p(X=0) = $\binom{7}{0}(0.30)^0(0.70)^7 = 0.082$. Notice that the combination symbol has appeared with k=0, so we're glad to have the definitions 0! = 1, P(n, 0) = 1, and $\binom{n}{0}$=1.

p(X $\geq$ 1) presents a small challenge, because X $\geq$ 1 means the formula must be computed for X=1, X=2, ... up to X=7. This isn't impossible or even extremely difficult, but it is annoying enough to make us wish for a shortcut. Fortunately, we anticipated this issue when we defined the Complements Rule. The complement of the event X $\geq$1 is the event X = 0, so we can use the complements rule:

0	**1**	**2**	**3**	**4**	**5**	**6**	**7**

Thus, p(X$\geq$1) = 1 – p(X=0) = 1 – 0.082 = 0.918. Be careful, though: we soon will encounter a problem in which both E and its complement are too complicated for their probabilities to be computed directly.

Example 19 presented us with 15 people taking emissions tests in which success counted as passing the test and p(passing) = 0.80. We were to compute p(X=15), p(X=0), and p(X $\leq$ 14).

p(X=15) = $\binom{15}{15}(0.80)^{15}(0.20)^0 = 0.80^{15} = 0.035$. It's not very likely that everyone will pass the emissions test, at least under these assumptions.

p(X=0) = $\binom{15}{0}(0.80)^0(0.20)^{15} = 3.27E-11$, which is approximately 0. It's extremely unlikely that everyone fails the emissions test.

As in Example 18, we next have a complicated event that calls for a shortcut, so we very earnestly hope the complements rule will work.

0	**1**	**2**	**3**	**4**	**5**	**6**	**7**	**8**	**9**	**10**	**11**	**12**	**13**	**14**	**15**

In this case, the complement of $X \leq 14$ is X=15, so we're in luck: $p(X \leq 14) = 1 - p(X=15) = 1 - 0.035 = 0.965$. It's quite probable that at most 14 people will pass the test.

Identity theft was the focus of Example 20. Success was defined as a verified case of identity theft (remember, success doesn't have to be an outcome of which we approve) and p(success) = 0.01. We were given five trials and asked for $p(X \geq 1)$, $p(X \leq 5)$, and $p(X = 2)$.

Whenever we get problems involving $\leq$ or $\geq$, we should be prepared to try the complements rule.

0	**1**	**2**	**3**	**4**	**5**

The complement of $X \geq 1$ is X=0,

so $p(X \geq 1) = 1 - p(X = 0) = 1 - \binom{5}{0}(0.01)^0(0.99)^5 = 0.99^5 = 0.951$

The complement of $X \leq 5$ is the empty set! The given event represents the set of all possible outcomes—the sample space. Hence $p(X \leq 5) = 1$, which the reader can verify, if (s)he wants, by computing

$$\binom{5}{0}(0.01)^0(0.99)^5 + \binom{5}{1}(0.01)^1(0.99)^4 + \binom{5}{2}(0.01)^2(0.99)^3 + \ldots + \binom{5}{5}(0.01)^5(0.99)^0.$$

The probability of getting exactly two outcomes, $p(X = 2)$, is $\binom{5}{2}(0.01)^2(0.99)^3 = 0.00097$, or 0.01.

Finally, example 21 presents us with n=500 travelers, and we are to compute probabilities of various numbers of them booking hotel rooms via the internet. Success is booking online, p(success) = 0.35, and our objectives are to report p(X=500), $p(X \geq 200)$, and $p(X < 300)$.

p(X=500) is, by now, routine. It's the probability that all 500 travelers book online. We expect its probability to be quite low; this is indeed the case:

$p(X=500) = \binom{500}{500}(0.35)^{500}(0.65)^0 = (0.35)^{500} \approx 0.000.$ In fact, there are more than 200 zeroes in the decimal expansion before the first significant digit, so our hunch was correct: this event has a vanishingly small probability.

The second question, $p(X \geq 200)$, seems to call for the complements rule because there are so many individual events involved: X takes on the 201 values from 0 to 200. There's a problem, though: the complement, $X < 200$, also involves hundreds of elements:

0	1	2	…	199	**200**	**201**	**202**	**203**	**204**	**…**	**497**	**498**	**499**	**500**

We expected this problem: the complement won't help us. Both the event in question and its complement are too complex to compute by direct methods. $p(X < 300)$ is just as bad.

There are three choices. We could bite the bullet and compute, by hand, all 201 terms in the binomial expansion for $p(X \geq 200)$. But that's a bad idea: it would take us forever, and fatigue from trying to calculate so many terms would almost certainly cause errors that would be difficult and time-consuming to spot and fix. We could also write a short computer program using, say, C++ or Java, involving a DO-loop that would do the work for us. That's not a bad idea, actually; we could adapt and reuse the program for other, similar problems. Statisticians ought to know how to write programs. We won't take the time to explain how this is done, however; that's for another book. (Some software, like Minitab, have options for computing these problems.) The approach we will take involves approximating this problem using the normal distribution, returning to a point we made long ago in Chapter 1 about the efficacy of the normal distribution in such cases.

First, though, a few more observations about the binomial distribution:

Just as normal distributions have means and standard deviations, so do binomials. The mean of a binomial random variable is the product np, and the standard deviation is $\sqrt{np(1-p)}$.

For the binomial random variable $X \sim b(n, p)$, $\mu = np$ and $\sigma = \sqrt{np(1-p)}$

Thus, for example, in Example 18 above, $X \sim b(7, 0.30)$, so $\mu = 7x0.30 = 2.1$ and
$\sigma = \sqrt{7x0.30x0.70} = 1.2$, rounded.
In Example 19, $X \sim b(15, 0.80)$, so $\mu = 15x0.80 = 12$ and
$\sigma = \sqrt{15x0.80x0.20} = 1.5$, rounded.

You've probably noticed that probabilities of "at least" and "at most" problems are usually bigger than probabilities for single events. That is because "at least" and "at most" are really compound problems, as opposed to single events that are computed with the equal sign. This can be surprising at first, but it's important to dispel the following common misperception.

Students often think that the probability that X equals the mean is greater than the probability that X equals any other value, because the binomials tend to be bell-shaped and mean=median=mode.

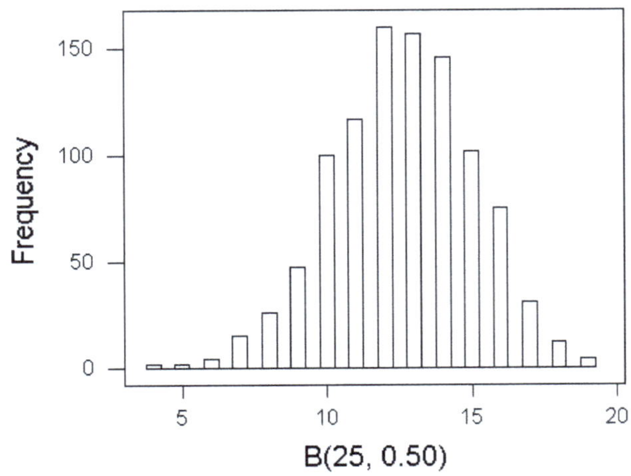

FIG. 4.10

The above graph shows 1,000 repetitions of a binomial random variable with n=25, p=0.50. The mean (or median, or mode) is 25x0.50, or 12.5. As the graph illustrates, the probability—or amount of area—of any particular outcome is very small. Much greater is the probability of a range of values. Even though 12.5 is the mean of X, this single outcome is improbable, but of course the area to the right of 12.5 is 0.50, as is the area to the left.

Finally, it's worth pointing out that complements need to be handled with care. Unlike continuous distributions such as the normal, for which the presence or absence of an equal sign does not affect the answer ($p(X < k) = p(X \leq k)$ and $p(X > k) = p(X \geq k)$), equal signs do matter for discrete distributions such as the binomial.

Suppose n = 5 and we desire $p(X \geq 1)$. The complement of $X \geq 1$ is $X < 1$, not $X \leq 1$. We could also write it as X = 0, of course. For discrete distributions, either the event or its complement can have an equal sign, but not both.

0	**1**	**2**	**3**	**4**	**5**

Similarly, if n=10 and we desire $p(X < 8)$, the complement of $X < 8$ is $X \geq 8$, not $X > 8$.

0	**1**	**2**	**3**	**4**	**5**	**6**	**7**	8	9	10

Also note that since discrete distributions depend on the number of occurrences of some event (for example, the binomial computes the probability of a specified number of successes), we must be ready to include the value 0. It's natural to start counting with 1, but this has to be questioned for discrete distributions.

Normal Approximation to the Binomial

As the above histogram suggests, binomials begin to act like normal distributions, approximately, under certain conditions. Thus, we can treat intractable binomials as approximate normal distributions, and we don't lose too much accuracy.

In an important sense, the interest of this approximation technique is mainly historical and pedagogical: historical because before the arrival of affordable desktop computing in the 1990s, this was the only practical way to perform these computations; and pedagogical because it demonstrates a relationship between these two distributions. As we mentioned in the first chapter, the normal distribution is important not because we find so many normal distributions in the real world, but because of the way the normal approximates other distributions. Anyone wishing to obtain the exact value of, say, $p(X < 300)$ when $X \sim b(500, 0.35)$ would probably use software. Nevertheless, the historical and pedagogical values of this technique are important enough for us to discuss for a few pages.

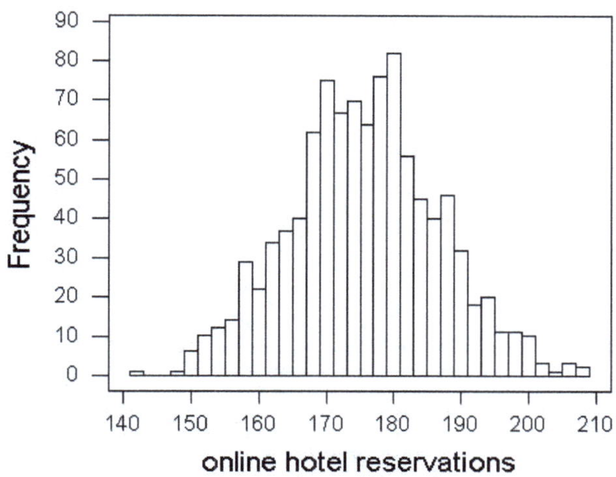

FIG. 4.11

Let's illustrate the technique using the data from Example 21. Minitab's graph of 1,000 samples from b(500, 0.35) is shown above. Notice that the mean is close to 500x0.35 = 175, and the standard deviation is $\sqrt{500x0.35x0.65} = 10.7$, rounded. Note also that the binomial data appear very close to bell-shaped.

To qualify for the normal approximation, we must first check that both $np > 5$ and $n(1\text{-}p) > 5$. We already found that $np = 175 > 5$; also, $n(1\text{-}p) = 500(0.65) = 325 > 5$. If both of these conditions are not met, this approximation technique should not be used.

The fundamental idea is the computation of the z score using the μ and σ computed above. As with the examples in Chapter 1, we will match the z score with a left-tail area, which we will obtain using Excel's **=normdist()** function, and make adjustments, as appropriate, depending on the area we desire.

Continuity Correction

There is one wrinkle, however, which the astute reader will have noticed: We're approximating a discrete distribution with a continuous one. There are bound to be rounding errors, and we need a technique for handling these.

The actual data under consideration are binomial, hence discrete. The approximating function is normal, hence continuous. The continuous curve will chop off the triangular corners of some of the rectangles or may include area between two rectangles, causing loss of precision.

The fix is called the continuity correction: It involves adding or subtracting 0.50 from the observation before calculating the z score. If the observation x is to be included in the shaded region, then add or subtract 0.50 to accomplish this, and similarly if x is not to be included.

Our current discussion calls for $x < 300$. Then we must shade to the left, but we do not want 300 to be in the shaded region ($X < 300$, not $X \leq 300$). Thus, we use 299.5 as the boundary of our probability region and compute the z score of this value:

$$z_{299.5} = \frac{299.5 - 175}{10.7} = 11.6, \text{ rounded.}$$

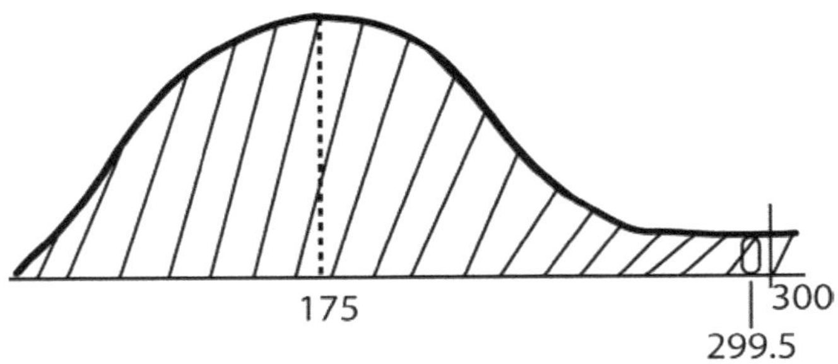

FIG. 4.12

The left-tail area is approximately 1 because the event in question is so much greater than the mean.

If we had wanted $X \leq 300$, we would have added 0.50 to obtain 300.5, so that when we shade to the left, we would shade through the value 300.

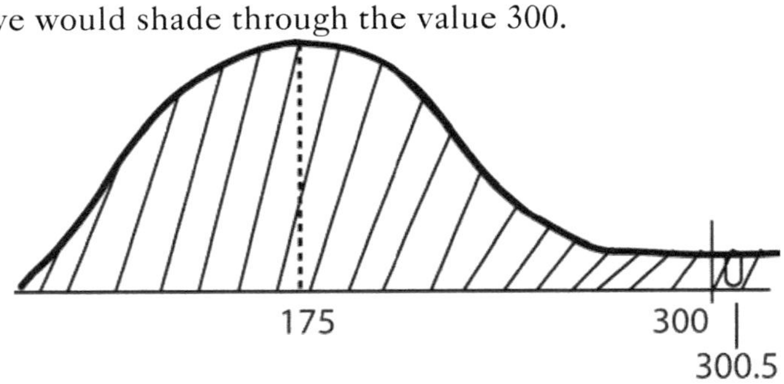

FIG. 4.13

To compute $p(X \geq 200)$, we would again employ the continuity correction, as follows: We shade to the right and want the number 200 to be included. Thus, subtract 0.50, using 199.5 as the boundary.

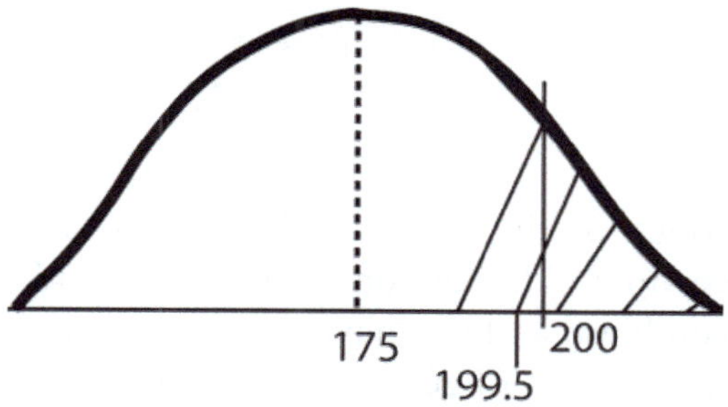

FIG. 4.14

$$z_{199.5} = \frac{199.5-175}{10.7} = 2.29\text{, rounded. The right-tail area is 0.011.}$$

Students sometimes find the continuity correction difficult to get used to, and some books include charts to determine whether to add or subtract 0.5. But the idea is really very simple and only includes these key ideas: always add or subtract 0.5, and this depends on whether the inequality includes an equal sign (that is, whether the boundary is or is not part of the shaded region) and the direction of shading. In my experience, close examination of only two or three examples is enough.

Also worth emphasizing is that the normal approximation has limited applicability. Sometimes students have the impression that all binomial distribution problems should be handled this way, having taken considerable time to derive and illustrate it; but that's not the case for two reasons. First, it is not always true that both $np > 5$ and $n(1-p) > 5$. Second, even when these two conditions hold, it could be that the problem can be solved directly or with the complements rule.

For example, the problems we have been discussing are based on $X \sim b(500, 0.35)$. Suppose we desire $p(X=0)$. This is easily computed as $\binom{500}{0}(0.35)^0(0.65)^{500} \approx 0$ without using the approximation despite the fact that the conditions have been met. Or suppose we are to compute $p(X \le 499)$. By the complements rule,

$p(X \le 499) = 1 - p(X=500) = 1 - \binom{500}{500}(0.35)^{500}(0.65)^0 \approx 0$, again without approximating.

Poisson Distribution with Mean μ

An extremely useful discrete distribution, called the Poisson with mean μ, is designed for computing the probability that a certain number of events occur in some measurement unit: hour, day, square foot, page, and so on. The events we have in mind are individually very unlikely, but we imagine very many of them: a large number of low-probability events.

Examples

Example 23

Suppose you're keeping track of the number of incoming text messages per hour. On average, you receive 4 texts/hour. The random variable X could count the number of texts you have received in a given hour, and we may want to compute the probability of getting no texts ($p(X = 0)$), or the probability of getting fewer than two texts in a random hour ($p(X < 2)$), or the probability of getting at least one text in a random hour ($p(X \geq 1)$).

The probability of getting any single text is quite small, but we should account for the possibility of getting a large number of texts.

Example 24

You might be interested in power outages per month in your neighborhood. It may be unlikely that you experience a loss of power in any given month, but we cannot rule out a large number of such events. If the average is 1 outage/month, and if X counts the number of such outages in a randomly chosen month, we may want to compute the probability of none ($p(X=0)$), or at least one ($p(X \geq 1)$).

Example 25

A certain weed sometimes appears in your garden. It appears very rarely, but you cannot discount the probability that it will grow out of control. If you divide your garden into one-square-foot plots and X counts the number of occurrences of the weed in a random square foot, and if you have observed an average of 3 weeds/square foot, we may wish to compute $p(X \geq 1)$, $p(X \leq 1)$, or $p(X=0)$.

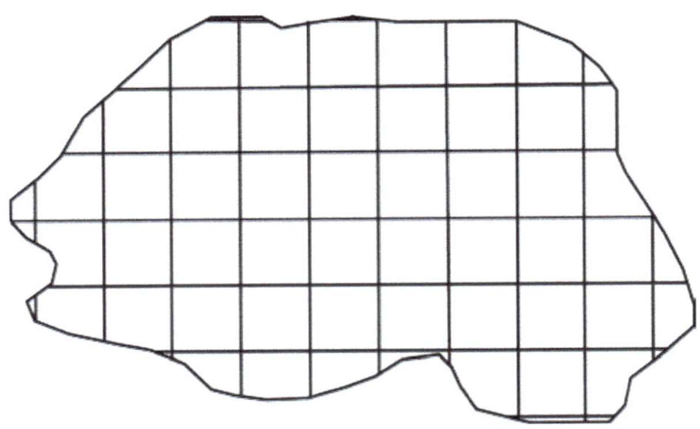

FIG. 4.15

The formula for the Poisson with mean μ is remarkably simple. It depends on μ; k!, the factorial of the number of successes; and e, the inverse natural logarithm of 1 ($e \approx 2.718$):

Poisson with Mean μ

If X is a Poisson random variable with mean μ that counts the number of occurrences of some event, then $p(X=k) = \frac{e^{-\mu}\mu^k}{k!}$, k = 0, 1, 2, …

Incidentally, the mean and variance of the Poisson are equal: mean = variance = μ.

Like the binomial distribution, the Poisson with mean μ is a discrete distribution that counts the number of occurrences of some event, although the Poisson does not consider these as successes and failures. But consider three important differences between them:

First, unlike the binomial distribution, no upper limit n is stipulated in the Poisson. The event in question can happen a very large number of times, and this number need not be stipulated in advance. The theory allows for it to happen infinitely often. In Example 23, there is no upper limit to the number of incoming texts; in Example 24, we do not stipulate in the phrasing of the problem how many power interruptions may happen; and in Example 25, you may hypothetically find an infinite number of weeds.

If there is no upper limit, how are we to calculate a problem like the last part of Example 23, the probability of getting at least one text? There will be an infinite number of terms! The simplest approach would be to use the complements rule:

$p(X \geq 1) = 1 - p(X = 0)$.

The ≥ problems in Examples 24 and 25 would be handled the same way. We will solve these problems shortly.

Second, the Poisson does not require the actual probability of the event being studied, other than that it is "very small." This is again unlike the binomial, in which p(success) must be stipulated.

Third, the Poisson uses a concept of independence that differs from the binomial. Recall that in the case of the binomial, two events E and F are independent if the joint probability equals the product of the individual probabilities: p(E and F) = p(E) x p(F). For the Poisson, independence refers to the size of the count per measurement unit: We assume, in the context of Example 25, that the number of weeds in any square foot does not affect the number of weeds in any other square foot. In Example 23, we are assuming that the number of incoming texts in any one hour does not affect the number in any other hour; and in Example 24, the number of power interruptions in any randomly chosen month does not affect the number in any other month.

Derivation of the Poisson from the Binomial

Before we turn to the solution of these problems, it is worth taking the time to derive the Poisson formula. What's important, and perhaps surprising, is that the Poisson is a special case of the binomial, under two assumptions: the number of trials n grows without bound, $n \to \infty$, and that at any given stage, the product np equals a constant, which we will label μ.

Let us assume, then, that we have an infinite sequence of binomial trials such that $n \to \infty$ and $p \downarrow 0$ in such a way that $np = \mu$, a constant. Warning: the last step in this discussion requires l'Hopital's Rule, which is taught in first-semester calculus. Readers who have not taken calculus or do not plan to take it will have to accept it on faith.

What we wish to evaluate, under these two assumptions, is

$$\lim_{n\to\infty} \binom{n}{k} p^k (1-p)^{n-k}$$

Since np=μ, we can rewrite this as

$$\lim_{n\to\infty} \binom{n}{k} \left(\frac{\mu}{n}\right)^k \left(1-\frac{u}{n}\right)^{n-k}$$

Let's expand the binomial coefficient $\binom{n}{k}$ as a permutation:

$$\lim_{n\to\infty} \frac{P(n,k)}{k!} \left(\frac{\mu}{n}\right)^k \left(1-\frac{\mu}{n}\right)^{n-k}$$

Expanding P(n, k) and using the exponents rules on the second and third terms (see Appendix 1 for a review of basic algebra), we obtain

$$\lim_{n\to\infty} \frac{n(n-1)(n-2)\ldots(n-k+1)}{k!} \frac{\mu^k}{n^k} \left(1-\frac{\mu}{n}\right)^{-k} \left(1-\frac{\mu}{n}\right)^n$$

Next, let us switch the denominators k! and n^k (again refer to Appendix 1):

$$\lim_{n\to\infty} \frac{n(n-1)(n-2)\ldots(n-k+1)}{n^k} \frac{\mu^k}{k!} \left(1-\frac{\mu}{n}\right)^{-k} \left(1-\frac{\mu}{n}\right)^n$$

Here's what all this algebraic rearranging has gotten us: In the first term, there are k factors n, n-1, n-2, … n-k+1 in the numerator and k terms, all n's, in the denominator. Each numerator factor can be paired with one of the ns in the denominator. The initial factor of n in the numerator cancels with one of the denominator ns, of course.

The second numerator factor, n-1, does not cancel with an n in the denominator. But remember that we're letting n increase without bound. In the limit as $n \to \infty$, the term n-1 becomes indistinguishable from n. Although there will always be some difference between them, this difference can be made as small as you like by choosing a big enough n. The same is true about n-2, n-3, and all the numerator terms down to n-k+1, when paired with the other denominator terms of n.

The concept of limit is discussed in detail in calculus courses, although some general-interest mathematics books introduce it.

The point is that by virtue of this limit argument we can cancel, pair-wise, n-1 with n, n-2 with n, and so on; the entire first term is 1. The previous expression simplifies to $\lim_{n\to\infty} \frac{\mu^k}{k!}(1-\frac{\mu}{n})^{-k}(1-\frac{\mu}{n})^n$.

The first fraction doesn't depend on n. k is a fixed constant that is unaffected by the increasing n, and μ, of course, we initially stipulated as the fixed constant np, hence we can write $\frac{\mu^k}{k!}\lim_{n\to\infty}(1-\frac{\mu}{n})^{-k}(1-\frac{\mu}{n})^n$.

The first term after the limit symbol, $(1-\frac{\mu}{n})^{-k}$, also approaches 1 in the $\lim_{n\to\infty}$ because of the n in the denominator and the fixed exponent k. Thus, we have reduced the original problem as follows:

$$\lim_{n\to\infty}\binom{n}{k}p^k(1-p)^{n-k} = \mu^k/k!\ \lim_{n\to\infty}(1-\mu/n)^n$$

This final limit expression cannot be evaluated by elementary operations, as in the previous steps. Here is where we need to call on l'Hopital's rule. It is proved in elementary calculus that

$\lim_{n\to\infty}(1-\mu/n)^n = e^{-\mu}$ where, as we mentioned above, $e \approx 2.718$ is $\ln^{-1}(1)$.

Calculus aside, what's most important for us is that we now have a closed expression:

$$\lim_{n\to\infty}\binom{n}{k}p^k(1-p)^{n-k} = \frac{e^{-\mu}\mu^k}{k!}$$

assuming $n\to\infty$ and $p\downarrow 0$ in such a way that $np = \mu$

This really is remarkable: The Poisson with mean μ and the binomial are special cases of each other. The Poisson is what the binomial turns into when there are very many small-probability events. Note especially how the condition that $n \to \infty$ matches our discussions in Examples 23, 24, and 25 about the lack of an upper limit for n.

Let's turn now to the solutions of those problems.

In Example 23, the distribution of text messages is Poisson with mean 4/hour. The probability of getting no texts in a given hour is

$$P(X = 0) = \frac{e^{-4}4^0}{0!} \approx 0.018$$

In a similar way,

$$P(X < 2) = P(X = 0) + P(X = 1) = \frac{e^{-4}4^0}{0!} + \frac{e^{-4}4^1}{1!} \approx 0.018 + 0.073 = 0.091$$

$$P(X \geq 1) = 1 - P(X = 0) = 1 - 0.018 = 0.982$$

Monthly service interruptions were discussed in Example 24. We assumed an average of 1 interruption per month and asked for $p(X=0)$ and $p(X \geq 1)$.

$$p(X=0) = \frac{e^{-1}1^0}{0!} = e^{-1} = 0.368$$

$$p(X \geq 1) = 1 - p(X = 0) = 1 - 0.368 = 0.632$$

In Example 25, we considered weeds in your garden per square foot. Assuming $\mu = 3$ weeds/square foot, we compute $p(X \geq 1)$, $p(X \leq 1)$, and $p(X=0)$ as follows:

$$p(X = 0) = \frac{e^{-3}3^0}{0!} = 0.05$$

$$p(X \leq 1) = p(X = 0) + p(X = 1) = 0.05 + \frac{e^{-3}3^1}{1!} = 0.05 + 0.149 = 0.199$$

$$p(X \geq 1) = 1 - p(X = 0) = 1 - 0.05 = 0.95$$

Multinomial Distribution

Six people apply for a position at a nearby company. Each applicant can be accepted, rejected, or waitlisted. The probability of being accepted is 0.20; rejected, 0.65; and waitlisted, 0.15. What is the probability that all six are accepted? That all six are waitlisted? That two are accepted, two are rejected, and two are waitlisted?

This problem seems to fit into the binomial framework, in that a maximum number of trials is stipulated (six) and individual probabilities are stipulated (0.20, 0.65, and 0.15). But there are three mutually exclusive and exhaustive categories, not two, hence the binomial will not apply. Perhaps we should say that the binomial will have to be adapted, and that is the focus of this section.

To this end, we'll rewrite the binomial in a way that suggests how to adapt it to a multinomial context. Start with the binomial formula:

$$p(X = k) = \binom{n}{k} p^k (1 - p)^{n-k}$$

The fractions p and 1-p are complements, hence sum to 1. Without loss of generality, let us rewrite $p = p_1$ and $1\text{-}p = p_2$ subject to the condition that $p_1 + p_2 = 1$.

Also, k counts the number of successes and n-k the failures. Again, without loss of generality, we can identify $k = k_1$ and $n\text{-}k = k_2$ subject to $k_1 + k_2 = n$.

Finally, let us rewrite the binomial coefficient:

$$\binom{n}{k} = \frac{P(n,k)}{k!} = \frac{n(n-1)(n-2)...(n-k+1)}{k!} = \frac{n(n-1)(n-2)...(n-k+1)}{k!} \times \frac{(n-k)!}{(n-k)!} = \frac{n!}{k!(n-k)!}$$

As I commented earlier, this last expression is how many textbooks define $\binom{n}{k}$. In my experience, it's easier to understand combinations as permutations divided by the factorial of the size of their subsets, but this (for us) new expression $\frac{n!}{k!(n-k)!}$ is very useful for introducing the multinomial, as we shall see.

Further, since $k = k_1$ and $n - k = k_2$, we can write $\binom{n}{k} = \frac{n!}{k_1!k_2!}$

What we have shown is that

$P(X = k) = \binom{n}{k} p^k (1-p)^{n-k} = \frac{n!}{k_1!k_2!} p_1^{k1} p_2^{k2}$ such that $p_1 + p_2 = 1$ and $k_1 + k_2 = n$.

This is exactly the sort of expression we had hoped for. Sparing you the mathematical details, for which you are invited to check the references, we know now how to generalize to the case of three, four, or more categories:

Multinomial Distribution with 3 Outcomes

$p(X_1 = k_1, X_2 = k_2, X_3 = k_3) = \frac{n!}{k_1!k_2!k_3!} p_1^{k1} p_2^{k2} p_3^{k3}$ where $p_1 + p_2 + p_3 = 1$ and $k_1+k_2+k_3 = n$

The generalization to four or more categories is obvious.

Let's solve the employment example:

We were given n = 6 and for each individual p(accept) = 0.20, p(reject) = 0.65, and p(waitlist) = 0.15.

Let's let X_1 count the number of applicants who are accepted, X_2 the number rejected, and X_3 the number waitlisted. Then, the probability that all six applicants are accepted is

$p(X_1 = 6, X_2 = 0, X_3) = \frac{6!}{6!0!0!}0.20^6 0.65^0 0.15^0 = 0.20^6 \approx 0.000$. Given the small probability that any individual is accepted, the joint probability that all six are accepted is very small.

The probability that all six are rejected is

$$p(X_1 = 0, X_2 = 6, X_3 = 0) = \frac{6!}{0!6!0!}0.20^0 0.65^6 0.15^0 = 0.65^6 = 0.075$$

The first two cases were easy, in fact almost trivial, because two of the three cases were zero, so the corresponding terms reduced to 1. The third problem, however, in which all three cases happen twice, takes somewhat more effort:

$$p(X_1 = 2, X_2 = 2, X_3 = 2) = \frac{6!}{2!2!2!}0.20^2 0.65^2 0.15^2 = 0.034.$$

The multinomial coefficient equals 90. It can be computed in stages or with Excel (6! = fact(6) and 2! = fact(2)). If you prefer to do it in stages, with pencil and paper, you could write 6! = 720, 2! = 2, so $\frac{6!}{2!2!2!} = \frac{720}{8} = 90$.

Contingency Tables

We close this section, this chapter, and this book by applying the ideas of independence, probability, expected value, and hypothesis tests—basically, all the major themes of this book—to the solution of a very interesting and important problem: tests of independence using contingency tables.

It's widely believed that obesity and diabetes occur together: individuals who are very much overweight are very likely also to have diabetes. But we are not interested in correlation, partly because we will not be using continuous data; rather, we are going to consider discrete data, or count-data, that have been cross-classified. For example, suppose we have 200 observations cross-classified as follows:

	Diabetes	No diabetes
Obese	80	20
Not obese	20	80

We will use this data to test the null hypothesis of no association versus the alternative that there is an association:

H_0: no association between diabetes and obesity
H_1: there is an association

As usual, we have stated a null hypothesis that we hope to reject. We imagine performing such a hypothesis test because we believe there is an association, so we collect a random sample, claim that there is no association, and hope to reject that claim. All hypothesis tests have this type of logic.

If there is no association, then if we randomly choose a participant in our study who is very much overweight, then there will only be a 50-50 chance of that person having diabetes. Equivalently, if there is no association, then a random participant known to have diabetes would only have a 50-50 chance of having diabetes.

To perform this test, we will need a test statistic, a critical value, and, of course, a distribution. If the test statistic exceeds the critical value, which is what we hope, we will reject H_0.

It can be shown (see the references) that the relevant distribution for tests of association with count data is the Chi Squared: the fourth distribution we have intensively studied (after the normal, t, and F). Excel will produce Chi Squared critical values using the function **=chiinv(probability, df)**. We turn now to the computation of the test statistic from the data.

The basic idea is this: Suppose H_0 were true. Then the counts we actually observed should very closely match what we would have expected under the independence assumption. The given table we will refer to as the table of observed counts. But how should we compute the expected counts?

The answer returns us to an important concept from Chapter 1: expected value. For contingency tables, the expected count in each cell is the probability that an observation is in the corresponding row and column, weighted (multiplied) by the total number in the experiment.

Observed			
	Diabetes	**No diabetes**	Totals
Obese	80	20	100
Not obese	20	80	100
Totals	100	100	200

Take the 80 participants in the upper left corner, those who are both obese and suffer from diabetes. The probability of having diabetes is 100/200 because there are 100 observations in the diabetes column, divided by 200 in the entire experiment. The probability that a participant is obese is also 100/200 because the total is 100 for the row labeled obese, also divided by 200. Thus, assuming independence, the joint probability is the product of the two individual probabilities: p(obese and diabetes) = p(obese) x p(diabetes) = 100/200 x 100/200.

What we really want, though, is not the joint probability; we want the expected count in that cell. Recalling our discussion in Chapter 1, an expectation is a weighted probability; in this case, the weight is the total n, 200.

$$\text{Expected count} = n \text{ x } p(\text{obese}) \text{ x } p(\text{diabetes}) = 200 \text{ x } 100/200 \text{ x } 100/200 = \frac{100 \times 100}{200} = 50$$

We will need to compute the expected count for each cell, but first I'd like to point out a feature of the above computation: For the upper left cell, $E = \frac{100 \times 100}{200} = \frac{r \times c}{n}$ where r stands for row total and c stands for column total. This formula for expected cell counts is quite common in the textbooks, and we will apply it here.

Here is the complete table of expected counts:

Expected			
	Diabetes	**No diabetes**	Totals
Obese	50	50	100
Not obese	50	50	100
Totals	100	100	200

If the expected count in each cell is very close to the observed count in the corresponding cell, then we will fail to reject the null hypothesis. But how do we measure these pairwise differences, and what is the threshold for rejecting/failing to reject H_0?

We will define the difference between each pair of cells as $\frac{(O-E)^2}{E}$ where O comes from the Observed table and E from the Expected table. It's called the Chi Squared difference. Then we will sum these expressions to obtain an omnibus test statistic for the entire experiment.

To start with the upper left cell, $\frac{(O-E)^2}{E} = \frac{(80-50)^2}{50} = 18$. You should convince yourself that all the Chi Squared differences will be 18.

Chi Square		
	Diabetes	**No diabetes**
Obese	18	18
Not obese	18	18

The omnibus or overall test statistic is the sum of these, or 72.

Is 72 big enough to reject the claim of no association? As always with hypothesis tests, we need a critical number, or threshold. We can get this from Excel's **=chiinv(probability, df)** function, remembering to enter 0.05 for probability (we would call it alpha); and we compute degrees of freedom as the product (#rows – 1) x (#columns – 1): for this problem, df = (2 – 1) x (2 – 1) = 1. Excel reports = chiinv(0.05, 1) = 3.841. Because the test statistic exceeds the critical value, we can reject H_0.

Notice that although this is a hypothesis test, this problem involved three elements that we first encountered in this chapter on discrete probability:

- we are analyzing count data instead of continuous data, specifically, tallies
- $p(E) = \frac{n(E)}{n(S)}$
- joint probability under the assumption of independence: p(E and F) = p(E) x p(F)

On the other hand, since it is a hypothesis test, we have a significance level (0.05 as the default, as always), a test statistic $\sum_{cells} \frac{(O-E)^2}{E}$, and a test statistic that we found from Excel. It's nice to close this book by coming full circle.

With one more example we'll be finished:

Let's imagine a test for the presence or absence of an association between dean's list and academic major. If we know someone's academic major, can we make a pretty good guess as to whether that person is, or is not, on the dean's list? Put differently, do some majors tend to issue higher grades than some others?

Observed				
	Math	**English**	**Computer Science**	Totals
Deans	**23**	**51**	**31**	105
Not Deans	**42**	**13**	**32**	87
Totals	65	64	63	192

We'll stipulate three majors, Math, English, and Computer Science, and of course either on dean's list or not on it. The 192 tallies in the Observed table are imagined to have come from a survey—they are actual, not calculated, tallies. One-hundred five participants are not on the dean's list; 87 are on it. Sixty-five respondents are Math majors, 64 are English majors, and 63 are Computer Science majors.

The entries in the Expected table are what the tallies would have been if H_0 were true, that is, if academic major were really independent of presence (or not) on the dean's list. Each cell tally is computed as $\frac{r \times c}{n}$. For the upper left cell, Math majors who are not on the dean's list, we compute $\frac{r \times c}{n} = \frac{105 \times 65}{192}$ = 35.5. Of course, we would never design a test if we thought the null were true; we test because we think it isn't true, and we want sample data to justify our rejection of the claim that it is true.

Expected				
	Math	**English**	**Computer Science**	Totals
Deans	**35.54688**	**35**	**34.45313**	**105**
Not Deans	**29.45313**	**29**	**28.54688**	**87**
Totals	**65**	**64**	**63**	**192**

There seem to be fairly big differences between corresponding cells in the Observed and Expected tables, though we would need to compute these as Chi Square differences using the formula $\frac{(O-E)^2}{E}$. For the upper left cell, $\frac{(O-E)^2}{E} = \frac{(23-35.5)^2}{35.5}$ ≈4.43. Let's get the other entries from Excel:

Chi Sq Differences			
	Math	**English**	**Computer Science**
Deans	**4.428633**	**7.314286**	**0.346096**
Not Deans	**5.344902**	**8.827586**	**0.417701**

The sum of these six differences, 26.7, is the test statistic for this problem. The critical number is =chiinv(0.95, 2) = 5.99, remembering that the Chi Square df is computed as (# rows – 1) x (# cols – 1) = (3 – 1) x (2 – 1) = 2.

Since the test statistic exceeds the critical number for this test, we reject the claim of independence between major and dean's list, which is the conclusion we hoped for. Knowing someone's major gives us a better-than-even chance of guessing whether that person is on the dean's list.

References

Lesch, S. and Daniel Jeske. (2009). "Some Suggestions for Teaching about the Normal Approximation to the Poisson and Binomial Distribution Functions." *The American Statistician*, 63, 274-277.

Minh, D. L. (2001). *Applied Probability Models*. Pacific Grove: Brooks/Cole.

Mirkin, B. (2001). "Eleven Ways to look at the chi-squared coefficient for contingency tables," *The American Statistician* 55, 111-120.

Proschan, M. (2008). "The Normal Approximation to the Binomial." *The American Statistician*, 62, 62-63.

Ross, K. (2004). *Discrete Mathematical Structures*, 5th ed. Upper Saddle River: Pearson Prentice Hall.

Ross. S. (1994). *A First Course in Probability*, 4th ed. Englewood Cliffs: Prentice Hall.

_______ (1993). *Introduction to Probability Models*, 5th ed. San Diego: Academic Press.

Rosen, K. (2003). *Discrete Mathematics and Its Applications*, 5th ed. Boston: McGraw-Hill.

Shader, M. (1989). "Two Rules of Thumb of the binomial distribution by the normal distribution." *The American Statistician*, 13, 23-24.

Snedecor, George W. and William G. Cochran (1989). *Statistical Methods*, 8th ed. Ames: Iowa State University Press.

Triola, Mario (2014). *Elementary Statistics*, 12th ed. Boston: Pearson.

Suggestions for Further Reading

Any good textbook on mathematical statistics will derive the mean and variance of the binomial and Poisson distributions. The derivations are not conceptually deep, but, as they involve expectation, there are lots of summations and changes of subscripts. For a beginning student, these details tend to obscure the main ideas, but the curious student might start with Ross (1994).

In addition to its relation to the Poisson, the binomial distribution can be adapted to account for variations in the way the successes are chosen or the way the trials are sequenced. See Lesch and Shader for discussions about the geometric, hypergeometric, and negative binomial distributions.

The Poisson distribution has far more interesting and important properties than we had time to discuss here. It plays a fundamental role in the development of more complex probability models, such as Markov chains and M/M/1 queues. See Ross (1994).

Bayesian statistics is vital for anyone who aspires to practice or even merely understand modern techniques. Unfortunately, most beginning textbooks skip Bayes' Rule or relegate it to appendices. We ourselves have barely scratched the surface, but see Ross (1993) and Rosen for further discussion. Be prepared, though: a full introduction to the Bayes paradigm is highly technical and calls on results from advanced calculus.

Chapter 4 Glossary

Bayes' rule	A special case of conditional probability used in this book on two-stage probability problems in which the condition applies to the second stage.
binomial distribution	If the random variable X has only two possible outcomes, each outcome has success probability p, there are n trials of which k have been designated successes ($0 \le k \le n$), and if the trials are independent, then $p(X = k) = \binom{n}{k} p^k (1-p)^{n-k}$
Chi Square difference	Used with contingency tables, the Chi Square difference between corresponding observed and expected cells is $\frac{(O-E)^2}{E}$. These differences are summed across all pairs in the contingency table to produce an overall test statistic for the test of association.
combination	If switching the relative positions of a subset of k objects chosen from a set of n is not counted as a separate arrangement, that is, if all that matters is the number of items in the subset and not the order in which they were chosen, then such an arrangement is called a combination of n objects taken k at a time and is computed C(n, k) = P(n, k)/k!
conditional probability	A condition is a rule by which the sample space is restricted .
contingency table	A matrix-type array of tally data for which the null hypothesis of independence between row and column is tested. The alternative is that there is an association between the row and column.
continuity correction	When using the normal distribution to approximate the binomial, add or subtract 0.5 from the observation value before computing the z score. Whether to add or subtract depends on both the direction of the inequality and whether the observation is part of the solution region or only its boundary.
expected count	Used with contingency tables: for each cell, the expected count is $\frac{r \times c}{n}$ where r is that cell's row total, c is the column total.
factorial	A special case of a permutation in which n = k, computed n! = n(n-1)(n-2) … 3(2)(1), and we define 0! = 1.
independence	For the binomial distribution, two events E and F are independent if p(E and F) = p(E) x p(F). For the Poisson, independence refers to the number of occurrences per measurement unit.

multinomial distribution	A generalization of the binomial: X_j can happen k_j ways, $1 \leq j \leq n$ where event j has success probability p_j
multiplication principle	The number of ways to complete a task is the product of the number of choices for the subtasks.
observed count	Used with contingency tables, the observed count in each cell is the tally of observations that match a particular row and column combination.
permutation	If switching the relative positions of a subset of k objects chosen from a set of n is counted as a separate arrangement, then such an arrangement is called a permutation of n objects taken k at a time, and it is computed P(n, k) = n(n-1)(n-2) … (n-k+1).
Poisson distribution	If there are a large number of low-probability events with mean μ, and X is the random variable that counts the number of outcomes, then $p(X = k) = \frac{e^{-\mu}\mu^k}{k!}$

Discrete Probability Review Problems

A. Bayes

1. In a certain city, the probability that a randomly chosen person has health insurance is 0.80. Among those who have health insurance, the probability that they will agree to lab work is 0.90; among those who do not have health insurance, the probability that they agree to lab work is 0.40.

a. What is the probability that a randomly chosen person will agree to lab work?
b. What is the probability that a randomly chosen person refuses lab work, given that that (s)he does not have insurance?
c. Compute the probability that someone has insurance, given that (s)he agrees to lab work.
d. Compute the probability that someone does not have insurance, given that (s)he refuses lab work.
e. Are questions c. and d. complementary events?

2. Along a certain stretch of highway near Chicago, suppose we classify drivers according to whether they are driving below the speed limit (15%), at the speed limit (60%), or exceeding the speed limit (25%). Of those driving below the limit, 70% are pulled over; of those at the speed limit, 20% are pulled over; of those going too fast, 80% are pulled over.

a. What proportion is pulled over?

b. What proportion is not pulled over? Do not use the complements rule.
c. What is the probability that someone was exceeding the speed limit, given that (s)he was not pulled over?
d. Compute the probability that someone was going at the correct speed, given that (s)he was pulled over.

B. Please compute these quantities:

a. $p(4,3)^2$	b. $p(p(3, 1), p(2, 1))!$	c. $3!^{2!}$
d. $\binom{3!}{2!}$	e. $p\left(\binom{4}{1}, \binom{3}{1}\right)$	f. $\binom{6}{3}^{2!}$
g. $\binom{p(6,1)}{p(5,1)}$	h. $\sqrt{\binom{9}{8}}$	i. $\binom{p(3,3)!}{\binom{2}{2}}^0$

C. Binomial and multinomial problems

a. Eight voters are waiting in line. The probability that any individual has a problem with his or her ballot is 0.09. What is the probability that everyone has a problem with his or her ballot? That no one does? That at least one person does? What is the mean number of ballot problems?

b. A certain corporation calls back 40% of its job applicants for a second interview. If there are 25 applicants, please compute the probability that everyone is called back for a second interview; that no one is called back; that at most 24 are called back. What is the mean number of people called back for a second interview?

c. A certain convenience store accepts cash, plastic, and phone payments. Suppose 20% pay by cash, 50% pay via phone, and the rest pay with plastic. If there are six people currently in line, please compute the probabilities that (i) everyone pays with a phone; (ii) everyone pays with plastic; (iii) two people pay via each method; (iv) one pays with cash, one pays with plastic, and the rest pay via phone.

D. Poisson problems

a. Suppose you receive, on average, four texts per day. What is the probability that on any given day, you will receive no texts? Exactly four texts? At least one text?

b. On your drive to work, you encounter, on average, five red lights. What is the probability that on any given drive to work, you will encounter no red lights? Exactly one red light? At least one red light? Is "exactly one" the complement of "at least one"?

E. Contingency tables

a. Is number of job offers independent of college major? Use the following table to test. Please state appropriate null and alternative hypotheses and report the Expected and Chi Square tables and the test statistic and critical value.

	Math	Accounting	English
1 job offer	44	67	31
2 job offers	27	56	22
3 job offers	19	30	22

b. Do men get more traffic tickets than women? Use the following table to test. Please state appropriate null and alternative hypotheses and report the Expected and Chi Square tables and the test statistic and critical value.

	Women	Men
1 ticket	55	80
2 tickets	25	80

Sample Final Exam #1

1. Suppose your conclusion from a One-Way ANOVA is to reject H_0. What would be the most appropriate next step in your analysis?

   a. There is no next step; you have all the information you need.
   b. Perform a post-hoc test such as the Tukey test.
   c. Compute $SS_{between}$ and SS_{within}.
   d. None of the above.

2. Suppose an unknown distribution has mean 110, standard deviation 14.5. Choose the true statement.

   a. Use the Empirical Rule to compute a confidence interval for μ.
   b. Use Chebyshev's Theorem to compute a confidence interval for μ.
   c. Since the distribution is unknown, it is impossible to compute a confidence interval for μ.
   d. The answer is not given by a., b., or c.

3. Which statement is always true regarding boxplots?

   a. $IQR > Q_3$
   b. $Max < Min$
   c. $Q_1 \leq Q_2 \leq Q_3$
   d. $Q_1 < Min$

4. Choose the false statement about normal curves.

   a. The y axis is an asymptote.
   b. Normal curves are symmetric around their means.
   c. Mean = Median = Mode.
   d. Approximately 0.000% of the area is beyond 4 standard deviations around μ.

5. If observation X is 2.75 standard deviations above the mean, then (choose the true statement):

   a. The distribution must be binomial
   b. mean = median = mode
   c. X must be an outlier
   d. $z_x = 2.75$

6. Which of the following is NOT a null hypothesis in Two-Way ANOVA?

a. Row and column are independent
b. Column means are equal
c. Row means are equal

7. Suppose we have the following Interactions Plot for a Two-Way ANOVA. What should be our next step in the analysis?

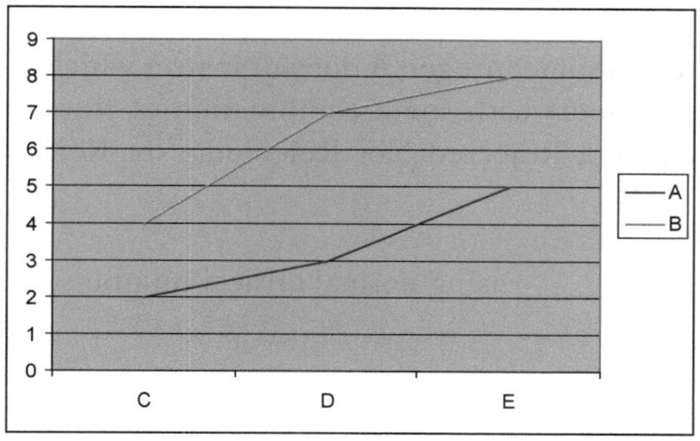

a. Stop the experiment, and do not analyze the main effects.
b. Do not analyze the main effects, but proceed to the Tukey post-hoc test.
c. Do not analyze the main effects, but proceed to the Bartlett-Levene test.
d. Analyze the main effects.

For problems 8–11, indicate the appropriate statistical model: Bayes, Binomial, Poisson, or Chi Squared (you do have to perform any calculations):

8. A vending machine eats two dollar bills per week, on average. The building manager wants to compute the probability that the machine eats at least one dollar bill in any given week.

9. An experimenter tallies the number of men and women who work zero jobs; one job; two jobs; three jobs. If the experimenter knows how many jobs a person works, can (s)he guess the person's gender?

10. The probability that any individual has health insurance is 0.45. Among 14 subjects, the experimenter wishes to compute the probability that at least half of them do not have health insurance.

11. 20% of respondents work two jobs. Of these, 15% are college graduates. Of those who do not work two jobs, 30% are college graduates. The experimenter wishes to know the probability that someone does not work two jobs, given that (s)he is not a college graduate.

12. **True or False:** *Please write the entire word TRUE or FALSE on your answer sheet.* Suppose a null hypothesis is rejected at 5% significance and two-tailed H_1.

a. A Type II error is possible
b. A Type I error is unavoidable
c. $p < \alpha$
d. The critical value exceeds the test statistic
e. The null hypothesis will also be rejected at 10% significance and two-tailed H_1
f. The null hypothesis will also be rejected at 5% significance and one-tailed H_1

For questions 13 and 14, suppose the z score for a given experiment is 1.97.

Assume H_0: $\mu = k$, where k is some integer, it doesn't matter which. Complete the following sentence (on the answer sheet) with some combination of the words Rejected and Not Rejected (Rejected-Rejected, Rejected-Not Rejected, Not Rejected-Not Rejected, Not Rejected-Rejected).

13. At $\alpha = 0.05$, H_0 will be ______ using normal critical numbers and _____ using t critical numbers.

14. Using normal critical numbers, H_0 will be ____ at $\alpha = 0.05$ and ____ at $\alpha = 0.01$.

15. Which one of the following is NOT an effect of increasing sample size?

a. Confidence intervals get shorter without sacrificing level of confidence
b. Critical numbers increase
c. Power increases
d. Prediction intervals for y that get shorter

16. Which one of the following quantities does NOT involve the standard deviation?

a. upper outlier boundary
b. z score
c. confidence interval for μ
d. coefficient of variation

17. Which one of the following quantities does NOT have 1 as an upper bound?

a. m
b. r
c. probability of an event
d. R^2

18. Suppose a certain regression has $s_x = 4.0$, $s_y = 5.0$, EV = 125, UV = 60. Fill in the blank: if the regressor increases by one unit, the response will change by ___ units, on average. (Assume a direct relationship.)

19. A One-Way ANOVA with four samples and five observations per sample has $MS_{between} = 30$, $MS_{within} = 5$. Please compute the effect size η^2.

20. Which one of the following techniques does NOT involve the assumption of normality?

a. regression
b. ANOVA
c. Contingency table
d. Bartlett's test for homoscedasticity

Sample Final Exam #2

I. Matching (one point each)

1. ANOVA	a. longer right tail
2. Contingency table	b. H_0: data are normal
3. Regression	c. $Q_3 - Q_1$
4. Measure of risk	d. H_o: μ's are equal
5. Normality test	e. C(n, k)
6. Right-skewed	f. H_0: row, column are independent
7. Default significance level	g. P_{50}
8. IQR	h. H_0: set of regressors is not useful
9. P(n, k)/k!	i. σ/μ
10. Median	j. 0.05

II. Power (10 points)

a. The power of a test is

i. The probability of correctly accepting a true H_0	ii. The probability of correctly rejecting a false H_0	iii. The probability of rejecting H_0 when it is true	iv. The probability of accepting H_0 when it is false

b. Which of the following can affect the power of a test?

i. sample size	ii. α	iii. the difference between the original and alternative μ's	iv. i, ii, and iii can affect power

III. 20 points

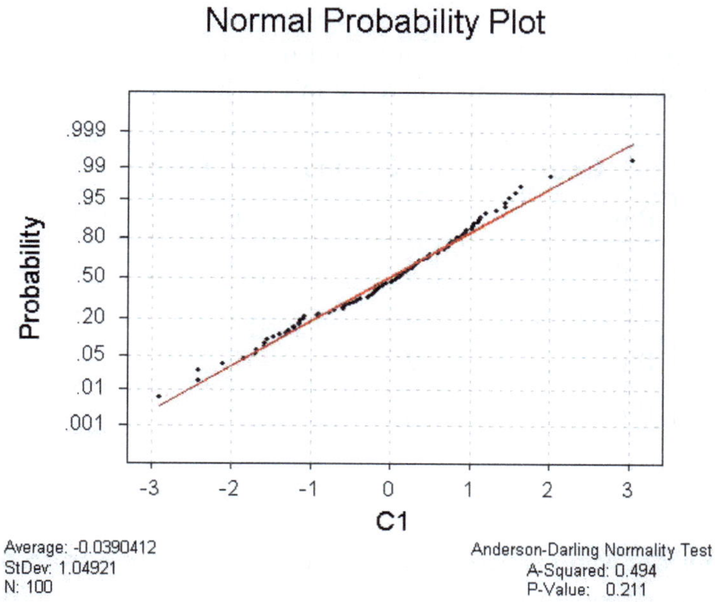

a. Approximately what is the mode of the data set?
b. Approximately what proportion of data lies between -3.18667 and 3.108589?
c. True or false: the data set probably has significant kurtosis.
d. True or false: the data set probably has significant left-skewness.
e. Considering the above data set as a random sample, please compute a 95% confidence interval for μ.

IV. 10 points

A study is conducted of traffic citations on a certain stretch of the I-88. The probability that any individual has gotten a speeding ticket is 0.35, a DUI is 0.15, improper lane usage is 0.20, and driving on a suspended license is 0.30. If four motorists are selected, please compute the probabilities that

a. none has a traffic citation
b. all have speeding tickets
c. there is one of each type of citation

V. 10 points

Please compute the student's GPA. Which would have the bigger effect on GPA: a one-grade increase in Physics or a one-grade increase in Psychology?

Course	Credits	Grade
Physics	4	B
Philosophy	3	A
Psychology	3	C
Photography	2	B
Podiatry	2	A

VI. 10 points

Suppose a certain body of water is being tested for the presence of a protected fish. Samples of water, each one cubic foot, are tested and reveal an average of 4.5 fish per cubic foot.

a. What is the probability of finding no fish in a random cubic foot?
b. What is the probability of finding at least one fish in a random cubic foot?
c. Please explain what the independence assumption means in this problem.

VII. 10 points

a. If order doesn't matter, please compute the number of ways of arranging seven books by topic, of which there are two: statistics (three books) and calculus (four books).

b. If order doesn't matter, please compute the number of ways of arranging seven books by topic, of which there are three: statistics (three books), calculus (two books) and philosophy (two books).

VIII. 10 points

Please NEATLY draw the interactions plot corresponding to this Two-Way ANOVA.

	A	B	C
D	5	4	3
	3	2	3
E	6	2	6
	4	1	7

IX. 10 points

In an experiment on factors that affect educational achievement, 45% of subjects participated in a review session before attempting a quiz and 55% did not. Of those who participated in the review session, 25% got an A on the quiz, 55% got a B, and 20% got a C; of those who did not have a review session, 15% got an A, 20% got a B, and 65% got a C.

a. What is the probability that a randomly chosen person got a B on the quiz?

b. What is the probability that a randomly chosen person got an A on the quiz, given that (s)he had the review session?

c. What is the probability that a randomly chosen person did not have a review session, given that s(he) got a C on the quiz?

X. 15 points

The data in the following contingency table are tallies of number of political events visited by the respondents. Columns represent gender and rows represent political affiliation.

	M	F
Democrat	50	60
Republican	40	50
Independent	30	30

a. Please state an appropriate null hypothesis.
b. Please compute the expected number of female Republicans who visit a political event.
c. Following is the table of X^2 differences. Please fill in the missing value (Male Independents) then state a conclusion. The critical number for this problem is 5.991.

	M	F
Democrat	0.01	0.01
Republican	0.06	0.05
Independent		0.16

XI. 20 points

Twenty subjects were asked to report the amount of minutes it took them to do 25 push-ups (y) as well as the average amount of minutes they spend per week on all other sorts of exercise (x) like swimming, running, weight lifting, and so on. That is, the experimenter wished to use general exercise to predict time to complete 25 push-ups.

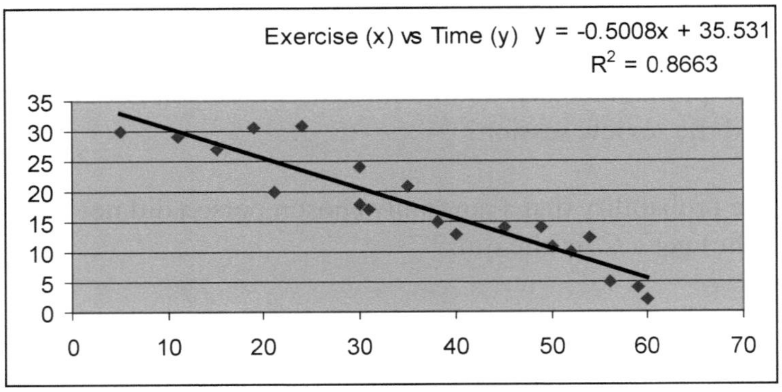

a. Please predict the number of minutes to complete 25 push-ups if a subject otherwise exercises 30 minutes per week.
b. What would be the effect on push-up completion time for one additional minute of overall exercise?
c. Please compute the correlation coefficient.
d. What proportion of Total Variation is explained by the regression?
e. Please report a 95% confidence interval for the population slope coefficient. You will need this information: TV = 1544.3, EV = 1337.57, $t_{18} = 2.101$, $\Sigma(x - \bar{x})^2 = 5333.2$. Will the experimenter be displeased with this result? Explain but be brief (no more than five words).

XII. 20 points

Source	SS	df	MS	F_{calc}
Between	1000			
Within		5		
Total	1500	25		

a. On your answer sheet, please report the missing quantities.
b. How many samples were in this experiment?
c. Please state the null hypothesis.
d. Please report your conclusion and possible error type.
e. Is Tukey called for? Explain.

XIII. 5 points

For hypothesis tests, the p value is ... (choose all correct answers)

a. 0.05
b. the probability of getting an outcome that is less likely than the outcome you actually got
c. the same as a Type I error
d. the same as the power
e. the boundary of the rejection region
f. the critical value

XIV. 5 points

For hypothesis tests, a Type I error is ... (choose all correct answers)

a. impossible if H_0 is rejected
b. certain to happen if H_o is accepted
c. the probability of correctly rejecting a false H_0
d. the same as the p value
e. the same as the significance level
f. the boundary of the rejection region

Sample Final Exam #3

1. (10 points) The following table gives a student's marks in a certain course (second row), as well as the point values. What grade would (s)he need on the final exam in order to earn an A in the course? Assume the lowest A is 90%.

T1/50	T2/75	T3/100	Final/200
44	67	85	x

2. (10 points) Which one(s) of the following conditions would probably cause a violation of normality?

a. if the mean were less than the standard deviation
b. if there were outliers
c. if the mode were much greater than the median
d. if the coefficient of variation were greater than 1.0
e. if the IQR were less than half the range

3. (10 points)

a. Regarding 95% confidence intervals for μ, please interpret the number 95%.
b. As sample size n increases, the length of the interval

i. increases
ii. decreases
iii. stays the same
iv. it is impossible to say in general

4. (10 points) Suppose H_0: $\mu=70$ versus H_1:$\mu\neq70$, and $\bar{x} = 72$, s = 9.05, n=81, tinv(0.05, 80) = 1.99, α=0.05. Please choose the correct statement:

a. H_0 will be rejected if critical numbers from the normal distribution are used, but not if t critical numbers are used
b. H_0 will be rejected if critical numbers from the t distribution are used, but not if normal critical numbers are used
c. H_0 will be rejected no matter whether critical numbers from the normal or t are used
d. H_0 will be not be rejected no matter whether critical numbers from the normal or the t are used

5. (30 points) Suppose a One-Way ANOVA is performed on four samples, A, B, C, and D. There are five observations in each sample. The computer output is reproduced here:

Source of Variation	SS	df	MS	F	P-value	F crit
Between Groups	6	3	2	1.333333	0.298523	3.238872
Within Groups	24	16	1.5			
Total	30	19				

a. Please state the null and alternative hypotheses.
b. What should be your conclusion?
c. Please draw the appropriate F curve, and label F_{calc}, $F_{critical}$, and the p value region.
d. What is the area under the curve to the right of 3.238872? Answer with a decimal.
e. What is the probability of getting an outcome more extreme than the one you actually got? Answer with a decimal.
f. Please compute the effect size η^2.
g. Is the Tukey post-hoc test called for? Why or why not?
h. What information is conveyed in the following graph? Be specific but brief.

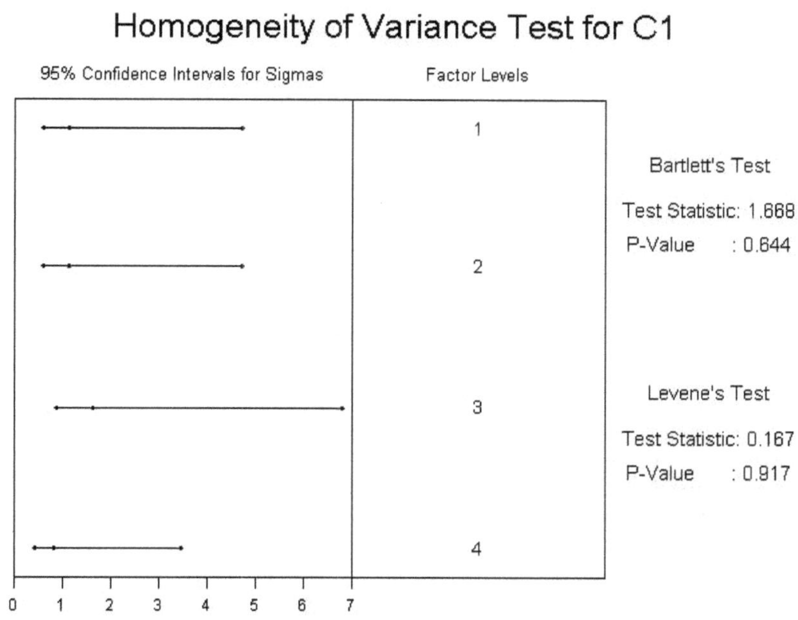

6. (15 points)

a. What information is conveyed in the following interactions plot? The context is Two-Factor ANOVA.
b. What does the graph imply about how or whether to proceed with the analysis?
c. As part of your answer, please define the word "interaction" in terms of interaction block means. Be as brief as possible.

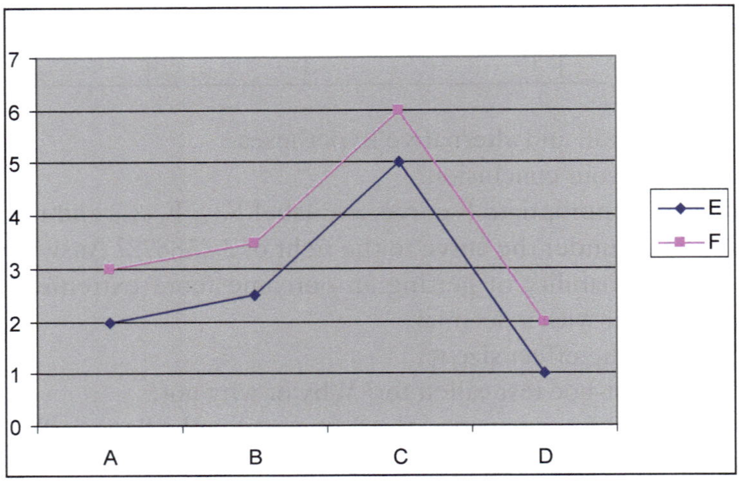

7. (5 points) Your friend, whose understanding of statistics is limited, is preparing a One-Way ANOVA on the following data. What should you tell him before he starts the analysis? Be specific but brief (five words or fewer).

A	B	C
3	7	5
5	4	5
4	2	3
6	5	82

8. (5 points) Next, your friend proposes a Two-Factor ANOVA analysis on the following data set. What should you tell him before he starts the analysis? Be specific but brief (five words or fewer).

	A	B	C	D
E	4 4	2 5	3 2	3 3
F	2	4	4	5

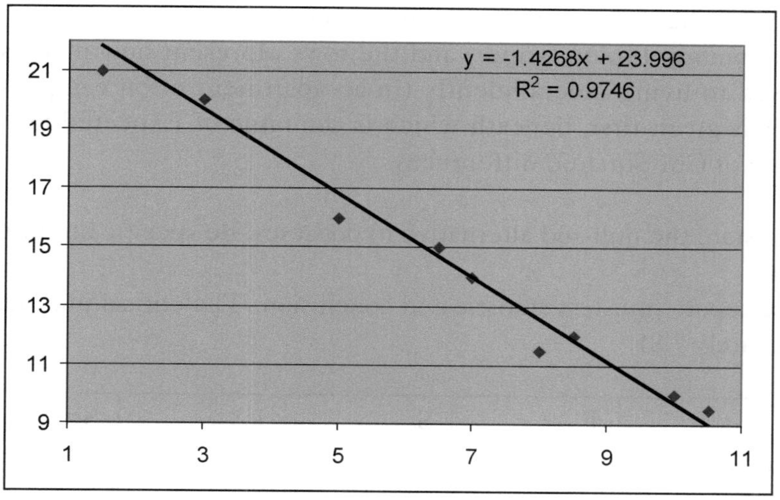

9. (30 points) A regression was performed on high school students to determine the effect of physical exercise on time to complete a quiz. Students were asked to report how many hours they spent per week exercising (regressor) versus the quiz completion time in minutes (response). The data were entered into Excel, and the regression results reported above.

a. Please predict the quiz completion time for someone who exercised six hours.
b. For this data set, TV = 195.525. What proportion of this quantity is explained by amount of exercise per week?
c. What is the correlation coefficient?
d. Is the correlation significant? The critical value is approximately 2.31.
e. What would be the effect on quiz completion time, on average, of one extra hour of exercise?

10. (10 points) For any individual, p(flu shot) = 0.35. If X is the number who get a flu shot and n=12, compute $p(X \geq 1)$.

11. (10 points) Suppose there are six visits per hour, on average, to your Facebook page. What is the probability that in any given hour, your page will have at least one visit?

12. (15 points) Let us say that 15% of motorists buy a new car battery this winter. Of these, 60% also will buy snow tires; of those who do not buy a new battery, 30% will buy snow tires.

a. What proportion of motorists will buy snow tires?
b. What is the probability that a random motorist will buy a new battery, given that (s)he will buy snow tires?

13. (20 points) In the following experiment, the columns represent year in college (Freshman, Sophomore, Junior, Senior) and the rows represent how many subjects live at home, as opposed to living independently (in an apartment or on campus). The table of Observed tallies is given first, beneath which is the table of Expected tallies and, on the bottom, the table of Chi-Squared differences.

a. Please state the null and alternative hypotheses. Be specific but brief.

b. Please report your test statistic and conclusion. The critical number for this test is approximately 7.81.

	F	S	J	Sr
At Home	50	40	30	25
Not at Home	25	35	45	65
E				
	F	S	J	Sr
At Home	34.52381	34.52381	34.52381	41.42857
Not at Home	40.47619	40.47619	40.47619	48.57143
X^2				
	F	S	J	Sr
At Home	6.937603	0.868637	0.592775	6.514778
Not at Home	5.917367	0.740896	0.505602	5.556723

Sample Final Exam #4

1. Which one(s) of the following quantities should an experimenter desire to be as large as possible?

a. probability of Type I error
b. probability of Type II error
c. power
d. p value

2. The Poisson distribution is a limiting case, as $n \to \infty$, of which one(s) of the following?

a. t distribution
b. f distribution
c. Chi Square distribution
d. binomial distribution

3. For binomial-type experiments, the independence assumption means that

a. p(E and F) = p(E) + p(F)
b. p(E and F) = p(E) – p(F)
c. p(E and F) = p(E) p(F)
d. the size of the tally in one measurement unit does not affect that in any other measurement unit

4. Suppose we have a multistage experiment and wish to compute p(A | B). Bayes' Theorem would be called for in case

a. B is in the second stage
b. B is in the first stage
c. A and B are independent
d. the second stage has more branches than the first stage

5. Choose the true statement(s) about contingency tables experiments.

a. The null hypothesis is that row and column are independent
b. df = (r-1)(c-1)
c. The expected count in each cell is what the tally would be if the null were true
d. The null is rejected if the calculated statistic is greater than the critical number

6. Choose the true statement(s) about bell-shaped curves.

a. The x axis is an asymptote
b. The area beneath the curve is 1
c. All bell-shaped curves are normal
d. mean=median=mode

7. To say that the t test is more conservative than the normal is to say that (choose all true statements)

a. normal critical numbers are bigger
b. z scores are always bigger than t scores
c. p values for the t test are larger than p values for the normal
d. none of the above

8. In regression, the slope coefficient m gives (choose all true statements)

a. the response value when the regressor equals zero
b. the square root of the correlation coefficient
c. the square root of the standard deviation
d. the amount of change in the response if the regressor decreases by one unit
e. none of the above
f. a and d

9. Which one of the following is not associated with a null hypothesis?

a. One-Way ANOVA
b. Contingency table
c. Bayes' Theorem
d. Test for interactions in Two-Way ANOVA

10. For the experiment associated with the above curve (choose all true statements)

a. A Type I error is possible
b. The p value is less than alpha
c. Power will increase as $n \to \infty$
d. The null hypothesis will be rejected

11. For the experiment associated with the above curve (choose all true statements)

a. 5.25 is the biggest value of the test statistic that would support the null hypothesis
b. 5.25 is the boundary of the rejection region
c. 9.75 is the p value
d. 9.75 is computed from the data, 5.25 is found in a distribution table

12. Which of the following statements is/are part of the Central Limit Theorem (choose all true statements)?

a. All samples have mean μ
b. All samples have standard deviation $\sigma / \sqrt{n}$
c. Sample means are normally distributed only if the population is normal
d. The standard deviation of the set of sample means is $\sigma / \sqrt{n}$

13. Three of the following quantities are always equal. Please choose the one that is not equal to the others.

- **a.** IQR
- **b.** median
- **c.** P_{50}
- **d.** Q_2

14. Suppose a distribution has max = 90, Q_3 = 50, Q_1 = 40. True or False: This distribution is probably non-normal.

Sample Final Exam #5

1. (14 pts) Based on this graph, please answer TRUE or FALSE to each of the following questions on your answer sheet.

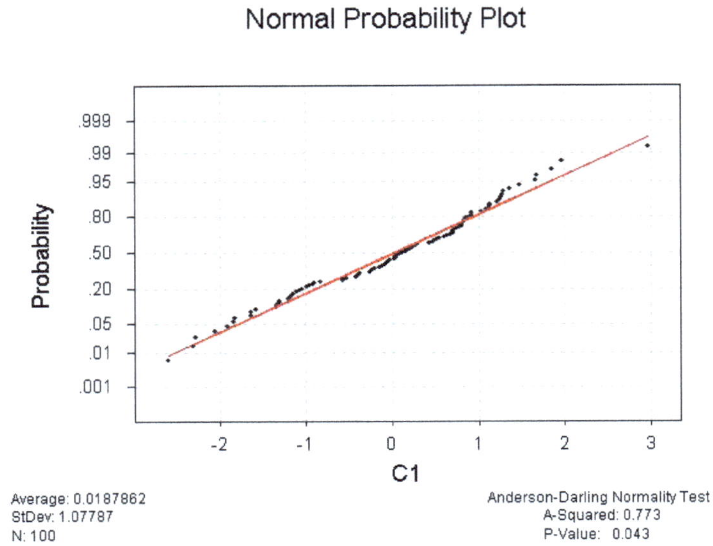

a. The standard deviation is probably negative.
b. The data set probably has significant skewness and/or kurtosis.
c. For this data set, Max < Min.
d. For this data set, Q3 + 1.5×IQR < Max.
e. For this data set, $Q_2 = P_{50}$.
f. This data set follows a normal distribution.
g. Probably, for this data set, median = mean = mode.
h. The mean probably has z score equal to zero.

2. (10)

a. A sample of 153 cell phone users revealed mean call durations 36.2 minutes, sample standard deviation 4.2 min. Please compute a 95% confidence interval for the true population mean call duration. You do not need to draw a curve.
b. Please interpret the number 95%.
c. Explain the trade-off, for fixed n, of confidence versus precision. Explain in no more than 10 words.
d. If n is not fixed, is there such a trade-off? Explain in no more than 10 words.

3. (10)

a. A sample of 12 flash drives revealed mean price $15.99, sample standard deviation 2.00. Please test H_0: $\mu=15.00$ versus H_1: $\mu\neq15.00$ at the default significance level. As part of your answer you must draw and label an appropriate curve, label your critical numbers, shade the p value region, and state the possible error type. The critical numbers are ± 2.179.

b. Please define the concept of p value in terms of probability using 15 words or fewer.

c. Explain why hypothesis tests based on the t distribution are "more conservative" than those based on the normal distribution. Use 10 words or fewer.

d. Was the Central Limit Theorem involved in your answer? Why or why not? If so, in what step(s) did you use it?

4. (10)

a. A normal population of weights of six-month-old kittens reveals $\mu=6.3$ ounces, $\sigma=0.77$ ounces. If there are 745 kittens in the population, how many of them weigh more than 10 ounces?

b. Did you use the Central Limit Theorem in your answer? Why or why not? If so, in what step(s) did you use it? Explain in 10 words or less.

5. (10)

a. It is desired to know whether student grades on a particular exam are affected by the time the course is offered (a.m., afternoon, p.m., weekend). Sample statistics report $SS_{Between} = 2070$, $SS_{Total} = 8519.429$, $df_{Between} = 3$, $df_{Total} = 27$, $F_{Critical} = 3.009$. Please report the ANOVA table for this experiment. Assume same-sized samples.

b. Please state the null and alternative hypotheses and your conclusion. Draw and label an appropriate curve and shade the p value region.

c. Is the Tukey procedure appropriate? Explain your answer in 10 words or less. You don't have to actually show any Tukey calculations, even if it is appropriate.

6. (5) This graph corresponds to a Two-Way ANOVA experiment. What does the graph tell you? Explain in 10 words or less. Be specific (but brief) and define the statistical terms you use. Assume there are two column treatment levels, A and B, and two row treatment levels, C and D.

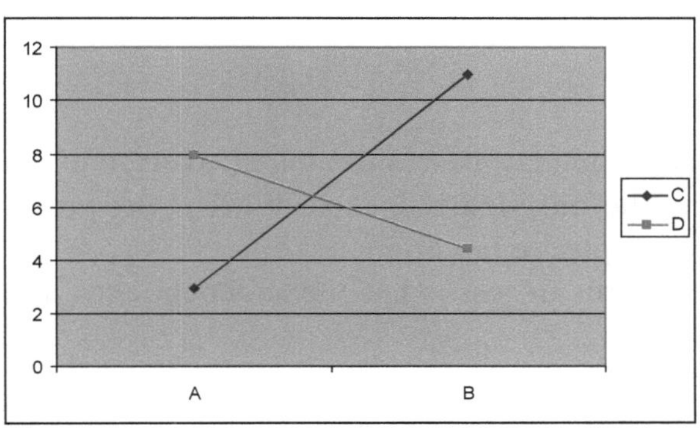

7. (10)

a. What would be the effect of a $1.00 increase in gas prices on hours spent watching television? Assume the regression equation is TELEVISION = 1.15*GAS + 0.45.
b. For this regression, assume EV = 200 and UV = 55.Compute the proportion of total variation explained by the regression. Compute the correlation coefficient. Assume a direct relationship.
c. Use its relationship with R^2 to compute F_{calc} assuming n=20 and k=1. Report the Regression ANOVA table.

8. (10)

a. Suppose X is a random variable with an unknown distribution. Please give a lower bound for the proportion of data that are within two standard deviations of the mean.
b. Repeat a. if it is known that X follows a normal distribution.

9. (6) Suppose data set A contains 20 gas prices from Lombard and B contains 20 gas prices from Aurora. Please answer True or False to the following questions:

a. The most appropriate statistical test based on these data sets would be regression, using A as the regressor and B as response.
b. The most appropriate statistical test would be regression, using B as the regressor and A as the response.
c. The most appropriate statistical test would be ANOVA, since A and B are independent samples, not matched pairs.

10. (10)

a. Toss three dice, and repeat for a total of six times. That is, toss three dice, then toss them again, then toss them again ... with the last toss of the dice being the sixth. Please compute the probability of getting a sum less than 18 on at most five of the tosses.
b. Same as part a., but compute the probability of getting a sum ***more than 18*** on at most five of the tosses.
c. Toss two coins. Please compute the probability of getting at most one head at least 145 times in 200 tosses.

11. (10)

a. At your company's holiday awards dinner, an average of five people at every table are texting during the president's speech. What is the probability that at any given table, at least two people are texting?
b. Explain, in 10 words or less, what the independence assumption means in this problem.

12. (5) A certain firm shows a profit of 25% for Year 2 (as compared to Year 1), 35% in Year 3, and 45% in Year 4. What single profit would produce the same result as do the individual profits?

13. (5) City A is 15 miles from B; B is 15 miles from C; C is 15 miles from D; and D is 15 miles from E. Suppose the only road connecting A and B has a speed limit of 35 mph; the only road connecting B with C has a limit of 55 mph; similarly for the road connecting C with D, whose limit is 40 mph; and the road connecting D with E has a limit of 50 mph. What would have to be that single speed such that a trip would take exactly as long as with the individual speeds?

14. (5) Suppose you save your holiday shopping until the last possible minute, and all the malls are crowded. Let X represent the probability (or proportion) of shopping malls with the number of available parking spots as indicated. Please compute the average number of available parking spots, as well as the standard deviation.

X	P(x)
0	0.30
1	0.35
2	0.20
3	0.15

15. (10) A pharmaceutical company is testing three drugs, A, B, and C, against a particular illness. Drug A is extremely effective on 75% of test subjects; mildly effective on 20% of its subjects; ineffective on 5% of its subjects. For drug B, the proportions are 60% extremely effective, 30% mildly effective, and 10% ineffective; for drug C, 25% extremely, 40% mildly, and 35% ineffective.

a. What is the probability that a randomly chosen subject shows no benefit after having taken a drug?
b. What is the probability that a randomly chosen subject was given drug B, given that (s)he showed a mild benefit?
c. What proportion of subjects showed extremely effective results?
d. Given that a subject showed no benefit, what is the probability that (s)he took drug C?

16. (10) The following data represent the number of job applications submitted in the past 30 days by job seekers, grouped by income (columns) and gender (rows).

	Income level (thousands of $)				
Gender		< 20	21-40	41-60	≥ 61
	M	31	54	55	43
	F	29	37	55	21

a. Please state an appropriate null hypothesis for this experiment.
b. How many men who earn between $41,000 and $60,000 would have submitted a job application if your null hypothesis in a. really were true?
c. Please report an appropriate measure of the difference between the observed and expected (under your null hypothesis) number of job applications submitted by women earning $61,000 or more.
d. The calculated test statistic for this experiment is 5.74 and the p value is 0.126. Based on this information, if you selected someone who submitted a job application in the last 30 days, what would you report (if anything) about her/his gender? Your answer must refer to your conclusion to the hypothesis test. Be brief (10 words or fewer).

Appendix 1: Mathematical Prerequisites

1. Rules of Exponents

a. $x^a x^b = x^{a+b}$

Base to a power times base to a power equals base written to the sum of the powers.

For example, $\sqrt{3} \times 27^{4/3} = 3^{1/2} \times (3^3)^{4/3} = 3^{1/2} \times 3^4 = 3^{1/2+4} = 3^{9/2}$

We used this rule in the derivation of the Poisson distribution from the binomial, specifically

$(1 - \mu/n)^{n-k} = (1 - \mu/n)^n(1 - \mu/n)^{-k}$

b. $\left(\frac{a}{b}\right)\left(\frac{c}{d}\right) = \left(\frac{a}{d}\right)\left(\frac{c}{b}\right)$

$$\left(\frac{2}{5}\right)\left(\frac{3}{7}\right) = \left(\frac{2}{7}\right)\left(\frac{3}{5}\right) = \frac{6}{35}$$

We also used this rule in the derivation of the Poisson, when we switched denominators k! and n^k.

2. Asymptotes

An asymptote is a line that a function approaches in the limit as a variable approaches some value. Asymptotes can be horizontal, vertical, or oblique (slanted), and some functions have more than one asymptote.

The function f(x) = y/x has a horizontal asymptote on the x axis and a vertical asymptote on the y axis.

3. Order of Operations

In your College Algebra or Precalculus course, you studied order of operations, probably focusing on nested parentheses: 3(x+2(4-x)) (which reduces to x+8). We are mainly interested in sums of squares versus the square of the sum.

Say our sequence of values is 2, 3, 5, 6, and 9.

The expression $(\sum x)^2$ means that we should first sum the values, then square this result. We know this because the exponent is outside the parentheses:

$$(\sum x)^2 = (2 + 3 + 5 + 6 + 9)^2 = 25^2 = 625$$

By contrast, sometimes we wish first to square the individual values, then sum these.

$$\sum x^2 = 2^2 + 3^2 + 5^2 + 6^2 + 9^2 = 155$$

Sometimes we take the square root of a sum of squares, but in this case the two operations do not cancel:

$$\sqrt{1^2 + 2^2 + 3^2} \neq 1 + 2 + 3$$

In fact, the left-hand expression equals $\sqrt{14}$, which certainly does not equal 6.

It is only permissible to cancel squares with a square root when the terms are being multiplied:

$$\sqrt{1\times 2\times 3} = 1\times 2\times 3 = 6.$$

4. Equations of Lines

In your previous courses, you probably studied several versions of linear equations, among them the general form ax + by=c and the slope intercept form y = mx + b. We are interested in the latter.

A line is a set of points whose coordinates, when substituted into the equation, give a true statement. For the line y = 2x – 1, the point (0,-1) belongs to the line as well as (1.4,

1.8) and (999.1, 1997.2). To spare ourselves the trouble, we typically graph a single point—the y intercept—then use the slope to find a second point. Since two points determine a line, this will be enough.

Appendix 2: Useful Excel Commands

=average()	the average of a column of data
=chiinv()	the critical number for the chi squared distribution
=combin()	the combination of n objects taken k at a time
=fact()	the factorial of the parameter
=finv()	the f boundary of a right tail beneath the f curve
=median()	the median of a column of data
=mode()	the mode of a column of data
=norminv()	the z score boundary of a left tail beneath the normal curve
=permut()	the permutation of n objects taken k at a time
=stdev()	the sample standard deviation
=stdevp()	the population standard deviation
=tinv()	the t score boundary of a left tail beneath the t curve

References

Balanda, K. and H. L. MacGillivray. "Kurtosis: A critical review." *The American Statistician* 42: 111-119.

Boslaugh, S. and Paul Andrew Watters (2008). *Statistics in a Nutshell.* Sebastopol: O'Reilly.

Casella, G. and Robert L. Berger (2002). *Statistical Inference.* Pacific Grove: Wadsworth.

Frankfurt-Nachmias, C. and Anna Leon-Guerrero (2015). *Social Statistics for a Diverse Society.* Los Angeles: Sage.

Hubbard, R. and M. J. Bayarri (2003). Confusion over measures of evidence (p's) versus errors (α's) in classical statistical testing. *The American Statistician* 7: 171-182.

Kennedy, P. (1994). *A Guide to Econometrics.* Cambridge: MIT Press.

Lesch, Scott M. and Daniel E. Jeshe. Some Suggestions for teaching about normal approximations to poisson and binomial distribution functions. *The American Statistician* 63: 274-277.

Lohr, S. (1999). *Sampling: Design and Analysis.* Pacific Grove: Duxbury.

Mosteller, F. and John W. Tukey (1977). *Data Analysis and Regression: A Second Course in Statistics.* Reading: Addison-Wesley.

Ross, S. (2010). *A First Course in Probability.* Upper Saddle River: Prentice Hall.

Saville, D. J. "Multiple Comparison Procedures: The Practical Solution." *The American Statistician* 44: 174-180.

Snedecor, G. and William G. Cochran (1989). *Statistical Methods.* Ames: Blackwell.

Stuart, A. (1984). *The Ideas of Sampling.* New York: Oxford University Press.

Sudman, S. (1976). *Applied Sampling.* San Diego: Academic Press.

Triola, Mario (2014). *Elementary Statistics.* Boston: Pearson.

Tukey, John W. (1977). *Exploratory Data Analysis.* Reading: Addison-Wesley.

9 781516 529117

Printed by Libri Plureos GmbH in Hamburg,
Germany